Under Water to Get
out of the Rain

Praise for *Under Water*

"To plunge into this book is to experience a glorious drenching. It is erudite, funny, weird and endearing. The deeps will never seem the same again." – *John Banville*

"Trevor Norton's beautifully written memoir of a life spent probing and pondering the sea depths derives much of its power from his observations ashore . . . What he saw is described with a novelist's sensibility and eye for detail . . . And his literate, witty, luminous prose makes this a marine biology to cuddle up to. This is a book to take to the seaside and to bed." – *Guardian*

"Lovers of the sea and sands will be swept away by *Under Water to Get Out of the Rain* in which the distinguished marine biologist Trevor Norton writes so lyrically that you can taste the salt of his bonding with the oceans, from the submarine lava tunnels of Lanzarote to the kelp forests of California and the pure silver strands of the Hebrides." –*The Times*

5/20 67 007.

Also by the same author

Stars Beneath the Sea
Reflections on a Summer Sea

TREVOR NORTON

Under Water to Get out of the Rain

A love affair with the sea

Illustrations by Win Norton

arrow books

Published by Arrow in 2006

3 5 7 9 10 8 6 4 2

Copyright © Trevor Norton 2005

First published in the United Kingdom in 2005 by Century
The Random House Group Limited
20 Vauxhall Bridge Road, London SW1V 2SA

Random House Australia (Pty) Limited
20 Alfred Street, Milsons Point, Sydney,
New South Wale 2061, Australia

Random House New Zealand Limited
18 Poland Road, Glenfield
Auckland 10, New Zealand

Random House South Africa (Pty) Limited
Isle of Houghton, Corner of Boundary Road & Carse O'Gowrie,
Houghton 2198, South Africa

Random House Publishers India Private Limited
301 World Trade Tower, Hotel Intercontinental Grand Complex,
Barakhamba Lane, New Delhi 110 001, India

The Random House Group Limited Reg. No. 954009

www.randomhouse.co.uk

ISBN 9780099446583 (from Jan 2007)
ISBN 0099446588

Typeset by SX Composing DTP, Rayleigh, Essex
Printed and bound in Great Britain by
Cox & Wyman Ltd, Reading, Berkshire

To Rachel and Paul, who have given me few worries but immense pleasure and pride, and of course Win, who made them possible.

Acknowledgements

I am for ever indebted to those people who inspired me to explore the margins of the sea. Hans and Lotte Hass I have acknowledged elsewhere. Sadly, it is now too late to thank my science master, Frank Graham, my research supervisor, Bunny Burrows, and my colleagues and mentors, Jack Kitching and John Ebling, but I would like to record my gratitude just the same.

Several people kindly supplied me with information and I am especially grateful to Professor Chuck Amsler, Dr David Anderson, Dr Einar Anderson, Dr David Bramwell, Lisa de Cesare, Dr Val Gerard, Dr Joanna Jones, Dr David Leighton, Dr Claes Lundgren, Professor Dan Morse, Professor Brian Moss, Dr Richard Nash, Dr Peter Neushul, Joan Parker, Professor Robert Waaland, the late Professor Wheeler North and, as always, Reg Vallintine. Roger Rawlcliff kindly checked my Latin and Greek.

Sincere thanks to Caroline Dawnay and Mark Booth for their unstinting encouragement and brave attempts to temper my worst excesses.

My wife, Win, has again produced lovely images where I had only words.

Preface

I grew up beside a sullen sea rimmed with coal dust, and that was where I first felt the tug of the tide. But the whisper in the shells was of bluer oceans beyond the horizon, salt-scented and transparent, alive with strange creatures.

So I searched for their shores and this is the story of my journey.

. . . it lies like sleep,
Like a big sea. Far off, far off, I feel the first wave . . .

Sylvia Plath

Credits

O'Faolain, S., *An Irish Journey*, Readers Union & Longmans Green, London, 1941. © 1940 by Sean O'Faolain. Reproduced by permission of the Estate of Sean O'Faolain, c/o Rogers, Coleridge & White Ltd, 20 Powis Mews, London W11 1JN

Plath, S. 1971, two lines from *Three Women a poem for three voices* published in *Winter Trees*, Faber & Faber, London, 1971. Reprinted by kind permission of Faber & Faber.

Ricketts, E.F., and Calvin, J., *Between Pacific Tides*, 3rd edn revised by J. Hedgepeth, Stanford University Press, 1952. Reprinted by kind permission of Stanford University Press.

Stark, F., *The Southern Gates of Arabia*, 1936. Reproduced by kind permission of John Murray Publishers.

Stephenson, T.A., *The British Sea Anemones*, Ray Society, 1935. The vignettes at the foot of page 153 is reprinted with permission from the Ray Society.

While every effort has been made to secure permissions, I apologise for any apparent negligence on my part and undertake to make any necessary corrections on future editions.

Seeing the Light

St Mary's Island, Northumberland

We lived on the second floor and viewed the sea through windows misted with salt. At night the beams from the lighthouse swept my ceiling.

I was seven when we had the worst winter for a century. The seaweed went stiff with rime and even the tide pools froze. A Greek freighter was driven ashore to perch upright and drip rust on to the rocks.

Snow drifted in an arc right up to my bedroom window and, if I'd had the courage, I could have slid down to the ground. Dad dug a tunnel out from the front door and carried me to school on his shoulders through a deep trench, and only I could see the surface of the sun-dazzled snow. For a few weeks we lived at the North Pole and I expected Father Christmas to sleigh round the corner at any moment. Then the myth melted, the streets turned to soiled slush and Santa, I suppose, was out of the question.

The following summer we walked to Curry's Point at the northern edge of Whitley Bay, where I lived. Two hundred years earlier I would have heard the clanking chains on the gibbet, where the corpse of a murderer, Michael Curry, hung until it disintegrated. From the Point was a causeway leading to St Mary's Island, where the lighthouse stood.

The 120-foot tower was the first really big thing I had ever seen close up, an immense rocket ship, white and wonderful. Inside, there were no cosy rooms with curved walls, just a hollow with a staircase spiralling around a central abyss. At the top, encased in a wilderness of prisms, was the lamp, like a crystal from another planet, able to incandesce and brush aside mere earthly darkness to explore my bedroom two miles away. The four-and-a-half-ton lamp floated on a lake of mercury and the slightest nudge from my finger would have made it revolve.

Lighthouses have shrunk a little since I grew, but I still love the big, clean, white ones, vertical and virginal yet audaciously erect, signals to sailors and signposts to God.

Whitley Bay was the last resort on the north-east coast of England. It squatted beside Newcastle, to which it was unnecessary to carry coal in those days. The words 'Whitley Bay' were on every visitor's lips as they sucked the sickly, lettered rock. Everything then seemed designed to damage your teeth: the rock, the candyfloss, scuffles outside the boozer. The cautious often removed their teeth as closing time approached.

The first time we went on holiday it was to Scarborough on the Yorkshire coast. As far as I could tell it was exactly the same as Whitley Bay, a land of leaky pails and broken spades and sandcastles that succumbed to the tide. I peered into the pools and collected red rubbery lumps in the hope they would blossom into anemones.

We stayed in a dishevelled hostel masquerading as a guest house. It was painted hospital white on the outside, boarding-house brown within. The landlady was firm but fair and had a face you could have abseiled down. It was cheaper if guests brought their own food for her to cook. You paid extra for a warm breakfast and a partial view of the sea, but you were locked out until teatime, even when it rained. We knew how to enjoy ourselves in those days.

＊

Although I didn't know it then, this was where the British love affair with the seaside had begun and in a most unlikely way. In 1660, a local doctor had published a book extolling the virtues of sea water for gout or 'drying up superfluous humours, and preserving from putrefaction . . . and all manner of worms'.

In France, medics forecast that sea bathing led to 'immediate death', but Scarborough's beaches became infested with scrofula sufferers and imaginary invalids, and bathing certainly lightened the load for those suffering from the effects of dirt. Later books extended the menu of afflictions amenable to salt-water cure to include 'ruptures, rheumatism and madness', as well as 'phrenzy and nymphomania'.

Bathers were indeed forcibly submerged by burly female 'dippers' – 'hideous amphibious animals', according to the artist Constable. The shock of those icy waters was fundamental to the cure, for it was well known that the 'terror and Surprize, very much contracts the nervous membrane and tubes, in which the aerial spirits are contained'. Scarborough assured its patrons that it had the coldest water of all.

In all senses the melancholia, worms and putrefaction were washed away with the tide. This gave nature's seaside sanatorium the edge over inland spas where, as Tobias Smollet feared, invalids with running sores might convert the warm baths into cauldrons of infection keen to dart into his open pores. Worse still, what if the bathwater got into the pump from which he drank the mineral water, and he was swallowing the 'sweat, dirt and dandruff and the abdominal discharges of various kinds, from twenty diseased bodies, parboiling in the kettle below'? No, the chill and voluminous sea was undisputedly a safer kettle of fish.

The 'beach' had been invented and many city dwellers would see the horizon for the first time. They came to enjoy the whipping of the waves which 'invigorates all parts', and a confrontation with the 'congenial horrors of unrestrained nature'. Others succumbed to a massage with freshly gathered seaweed, or simply admired the bathing beauties from afar while simultaneously celebrating the invention of opera glasses.

The lure of the seaside was that people were liberated to let their hair down in public, both metaphorically and literally. After bathing, a woman's wet tresses had to be brushed and the beach was the only place where they might be viewed free from the nets and buns in which they were normally imprisoned.

As soon as the 'lower class' arrived, they seemed to 'give up their decorum with their rail ticket and to adopt practices which at home they would shudder to even read of'. Indeed, 'men gambol about in a complete state of nature' and women frolic with only 'apologies for covering'. In 1866, the *Scarborough Gazette* contained angry letters claiming that a healthy recreation had been turned into 'an immoral and depraved exhibition'. On the other hand, traders knew from experience that 'if first-class visitors are obliged to wear drawers when bathing . . . Scarborough will lose its fame'. New by-laws divided up the beach into bedrawed and knickerless sections.

Scarborough Spa prospered. The visitors' book read like a Who's Who of high society. The town published a weekly gazette to list all the important new arrivals. It was the first to have bathing machines and had forty of them by the 1780s. Arcades, a covered promenade and assembly rooms were built to occupy the bathers when they were out of the water or if, inevitably, it was raining. Seaboard towns all over Britain followed Scarborough's lead, each boasting saltier water than the others, or younger female bathing attendants, or fewer of those wicked waves that 'annoy, frighten and spatter bathers exceedingly'.

In the 1850s you couldn't step on the English coast without

tripping over a resort bulging with 'attractions'. Visitors were deafened by the noise of barrel organs and brass bands.

But by the 1950s, when I came, even the grand buildings of Scarborough were about to fall on hard times. The desire to retire to the seaside to await death gave some places a bad name. Deadly Llandudno had the highest death rate in Britain.

In Whitley Bay, most of the iron railings had been melted down during the war and dropped on Hitler. The few that remained received their annual overcoat of royal-blue gloss, covering an undercoat of rust.

At the Spanish City there was still some of the fun of the fair: the Waltzer, dodgems smelling of sparks, and lairy lads in kiss-me-quick hats who stood on the cars chatting up your girl while taking the fare. In the amusement arcades, ball bearings clattered around in the ancient mechanical slot machines. Men in short white coats roamed the floor like disconsolate dentists. 'Change!' they cried. 'Any change required?'

The times they *were* a-changing. The gypsy palmist, having read between the lines, packed up and left.

I didn't misspend my youth; in fact, I didn't spend it at all. It just fell through a hole in my pocket. In summer my mates and I lounged on the beach listening to the shrieks of timid bathers, their skins in summer scarlet. The donkeys had been temporarily displaced by rides on an army surplus DUKW from which dads might relive the D-Day landings. We eyed the girls in their new nylon bathing cossies, which were a big advance on the old woollen ones as they were transparent when wet. On quiet days we amused ourselves by divining the character and lifestyle of those who had left their bum prints on the beach. The smallest pleat in the sand or asymmetry of cheeks would reveal acres of biographical detail. It was a form of palmistry and about as reliable.

The opaque water slithered in, then oozed out again. It never seemed to smile and it never got warm. Autumn always came too soon, with winds that caused the gull droppings to fall obliquely and congeal before they hit you.

I didn't enjoy being young. Perhaps I wasn't very good at it. My parents were summoned by the headmaster to be told that I was the worst boy in the school. To me this became almost a point of honour.

My essays, though lively, were decorated with extraneous scrawls and blots, which I rather liked, but they elicited tetchy comments such as 'Keep this book away from the dog!' Perhaps I was concerned that, should my handwriting improve, they might discover how little I knew. I came a creditable bottom of the class in woodwork: '15th out of 15 – Much improved work this term.'

Television was to change my life for ever with a series of films in which Hans Hass went *Diving to Adventure*. Each week he swam with giant manta rays and moray eels, and his luscious wife, Lotte perched on mounds of coral while sharks circled ominously. She looked good enough to eat.

In one film I'm almost sure she peeled and swallowed a banana underwater. I forget the biological significance of this, but it made a lasting impression on at least one fourteen-year-old boy. I coveted Hass's beautiful yacht and the freedom to rummage below on tropical reefs. And then, of course, there was Lotte. I decided that I too would become a handsome Austrian adventurer.

As soon as the spring sea warmed to almost frigid, I submerged beneath the waves for the first time equipped only with a pair of bathing trunks. Instead of the ripple of sun flecks on the pale sand, all I saw was a blur swilling below me, and a fuzzy cloud as a flatfish fled. Our eyes don't work in water. They need a layer of air to allow them to focus. So I bought a diving mask and a snorkel with a caged ping-pong ball to keep out the water. Begoggled, amphibian-footed and

carrying a toasting fork fixed to a bamboo pole, I slid off the
sandstone slabs at St Mary's Island and into the cold and cloudy sea.

In a kelp-lined gully the tall plants swayed back and forth in the
wash of the waves. It felt as if I were passing between rows of galley
slaves pulling against a big sea. I was fighting the swell one moment,
surging forward into a fizz of bubbles the next.

Bursting for breath, I thrust my head into the gloom beneath the
canopy of kelp. With each movement above there was a flicker of sun
and shadow on the rocks beneath. Then two seaweeds closed around
my neck with the soft hands of a strangler. I shot to the surface and
tried to look as unconcerned as Hans Hass in a swirl of sharks.

From the surface I could see a large dome about fifteen feet
below. I took a deep breath, folded my body at the waist then lifted
my legs out of the water. Their weight caused me to slide effortlessly
down. The mask squashed against my face and it felt as if a needle was
being inserted into my ears. The water was closing in on me.
Everything with air in it – my lungs, sinuses and ears – was being
compressed. Even a foot below the surface the compression on my
chest was equivalent to a weight of over 180 pounds. It was the same
on my abdomen, pushing the diaphragm up into the chest cavity and
squeezing it still further. Three feet underwater the pressure differ-
ential is so great that it is impossible to suck down air through a tube
from the surface, but I soon learned that snorting into the mask
pushed it out again, and swallowing hard took care of my ears.

It was a painful lesson, but I couldn't be distracted. Being underwater was more exciting than I had ever imagined. A kaleidoscope of
new images overwhelmed me: the elegant untidiness of lazily swaying
seaweed, the uncontrived encounters with silver sand eels and crusty
crabs. I had walked through some woods without seeing a single
squirrel or badger, but here wild animals came out to meet me.

On the bottom, the dome I had seen turned out to be a ship's rusty boiler and I stared into its dark and dangerous interior. Who knew what might be lurking inside? In the surrounding sand, softly lifting and settling in the swell, I found a rudder and a brass propeller with its shaft. It was my first wreck and it had waited sixty years for me to find it. I surfaced breathless, more from excitement than lack of air.

I went down again and again. Suddenly, while I was below, something stabbed the water in front of me, a dark javelin in a cone of bubbles. It transformed into a cormorant. Having misfished for a sand eel, it escaped back to the sky. It was the most wonderful thing I had ever seen and I have never seen it since.

This, I decided, was the real world. The air-bound attic up there beyond the surface was no place to live. This fresh and alive sea was everything that the land wasn't. We plod around on land, victims of gravity. It is merely a surface on which to stand, the wind a mischievous nuisance. But underwater, weightless and often powerless in the current, you become one with the flow.

On my final return to the surface a fluttering shoal of pollack parted to let me pass. They were neither anxious nor curious. I was just one of the boys. It was as if the sea had been expecting me.

Sitting on the rocks and trying to dry myself in the wind, I watched a heron pluck green crabs from the pools then soar away like a tired pterodactyl. Flights of knots and oystercatchers came in with the tide to feed or to loiter on one leg. It was a super place to shiver. There were a couple of cormorants chatting on a rock. Perhaps one was *my* cormorant, boasting to his chum about the one that got away: 'I saw the queerest thing today. Put me right off my fishing. So *ugly*. And it couldn't dive for toffee.'

The next day I sawed off the end of my snorkel and threw away the ping-pong ball. I would never again mind the taste of the sea. After all, the world was seven-tenths salty water and so was I, and the chemical composition of my blood was almost identical to that of sea water. The ocean was truly in my veins and briefly, when still in the womb, I even had gill slits.

On the Bottom

Whitley Bay, Northumberland

Everyone was shocked. It was completely unexpected. I began to pass exams.

So far I had promised little and lived down to everyone's expectations. Secondary-modern kids weren't even supposed to take exams. We were on the bottom of the educational ocean.

French might still be a problem. My French mistress prophesied that by the time I learned to speak the language I would be too old to cross the Channel.

My parents thought a little coaching might help, so they exhumed from the most Yellowed Pages Miss de la Motte, who would give me an hour every Wednesday evening during the winter. She lived in a musty villa with a dark staircase leading to locked attic rooms. The parlour was laden with the past. It was as if a survivor of the Tsar's family had fetched up here with a wagonload of furniture and photographs. There were heavy velvet drapes redolent of damp decay, fat candles and guttered volcanoes of wax everywhere, extinct gas brackets and an old bell-push on the wall. Long ago, the ivory button had summoned servants, now it dangled on a wire like an escaped eyeball.

Every surface was dense with dust. Not the friendly fluff that hides under the bed, but the deep silt that accumulates on the bottom of an unvisited ocean.

The chill and the gloom added to the feeling that we were trapped in a wrecked ship lost in the abyss. At any moment an octopus might slither out from behind the curtains.

Miss de la Motte lived alone. Her face was as crumpled as discarded paper, but she had wonderful porcelain-blue eyes. Great whorls of hair gathered on her head like a beret after rain.

Over the empty fireplace was a painting of a placid stream. On the bank, half hidden in the shadows, sat a wistful young girl in an Edwardian apron and dress. She had cast a stone into the water and was watching the ripples expand, wondering perhaps where she might travel to and who she might become. The girl had porcelain-blue eyes.

I can still hear Miss de la Motte's voice, that rich whisper of an ancient record, punctuated by the soft rattle of loose teeth. Yet she spoke more perfectly than anyone else I have known. Although she never dusted an external surface, no sentence emerged until it had been polished for public view.

We used an ancient textbook and however carefully I opened its desiccated maroon covers, it creaked like the door to a haunted house. In its corridors lurked an encyclopaedia of unknown words, hiding beneath grave or acute eyebrows and circumflex frowns. I can't recall the lessons except for the phrase *l'enfant refractaire*. As with all the other phrases, I had to look it up in the glossary. It said: 'the refractory child'. I was none the wiser.

There were only so many frozen Wednesday nights that I could stand. I had better learn the language, and amazingly I did. One day the *sous* dropped and French didn't seem so hard. I passed the exam and began to rise towards the surface.

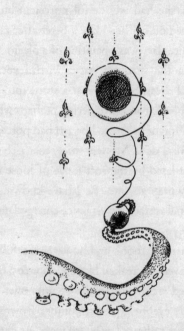

Changes

Collywell Bay, Northumberland

Just up the coast, flying a pennant of smoke, was Blyth. Cambois, nearby, was pronounced 'Camuss' to prevent the inhabitants getting ideas beyond their station.

The taste of coal was on the wind and the town rang with the clank and rumble of shunting wagons and rock tumbling down the loading chutes into echoing ships' holds. Everything was the colour of anthracite: the docks, the dockers and the rusty coasters. More coal was being shipped from this little port than from any place in Europe.

I transferred to Blyth Grammar School. Each day I travelled up by train and each day it was late because, they said, the weather was too cold (frozen points), too hot (buckled points), or too wet (rusted points). Once, if the newspaper was to be believed, the delay was a result of the driver being too short to reach the pedals.

The suburban train to school was one of those without corridors, just a series of separate compartments. When you stepped into that enclosed box you never knew who might await you; a leggy blonde with Michelin lips, or maybe a mad axeman – but it was usually a shambles of untidy school kids.

The journey to school passed through mining country. The footballing Charlton brothers were born just north of here in

Ashington and had they stayed they would probably now be dead from pneumoconiosis. Everywhere there were conical spoil heaps like Fujiyama foothills. They remained resolutely black, shunned by vegetation. Someone had the idea of planting their slopes with cherry trees to help them 'blend in', but the sad saplings failed to thrive. Ashington was once the biggest mining town in the world, but twenty-one years after I first saw it, the last mine would close.

I studied geology, zoology and botany, and games were of course compulsory for everyone except the games master. Like all games masters, ours limped around in an elastic knee bandage, which mysteriously changed knees. It prevented him from joining our football match in the mud and meant he had to skip the gruelling cross-country runs. Luckily, his lungs were sound so he could shout a lot and smoke energetically while we ran around in the rain. So I learned to play football and studied a little – much too little.

A new biology teacher encouraged us to do field projects, so I headed for the shore. Sand dunes stretched south from Blyth and the nearest rocky shores were beside Seaton Sluice. It isn't much of a name for a place, nor much of a place for a name, but it had once been impressive. From the thirteenth century, salt had been manufactured here by evaporating sea water in huge pans over fires. By the eighteenth century, large amounts of coal were being shipped from here to London. An artificial harbour was built for Sir Ralph Delaval and served twenty-two resident ships. Outside the harbour there was still an island composed entirely of Kentish chalk dumped as ballast long ago. But the harbour silted up and, to make a more direct route to the sea, a sluice was cut through the rock, almost fifty feet deep, thirty wide and nine hundred long. Gates on the seaward end once opened and closed automatically with the tide.

Seaton Sluice prospered. The Delavals set up a glass factory with expert glass blowers imported from Germany. By 1777 it was the largest glassworks in Britain, producing one and three-quarter million bottles a year. Around the tiny harbour were seven alehouses to refresh the workers.

The Delavals required a residence to reflect their station in life, so they called in a playwright. John Vanbrugh had 'a heart above his income' and took only six weeks to pen his first play. It was a hit, as was his next effort. When a rich dilettante decided to build a new theatre in London, who better to design it than a theatrical fellow like Vanbrugh. Unfortunately, the grandiose temple he constructed had the acoustics of an eiderdown.

When next invited to design a building he asked Nicholas Hawksmoor, the chief draughtsman of St Paul's, how it was done, then went off and did it. The result was Blenheim Palace. Next was Castle Howard, but he found Yorkshire 'so bloody cold, I have almost a mind to marry to keep myself warm'. There were candidates aplenty for love was 'as much forced up here as melons'. An observer wrote that 'His inclinations to ruins has given him a fancy for Mrs. Yarborough'.

Vanbrugh's next project lay a mile upstream from Seaton Sluice where he built a Palladian palace as confident as the Delavals themselves. The house was rumoured to have had tilting beds that tipped guests into a cold bath, and sliding walls to reveal them in bed with a serving wench.

To turn into the courtyard at Delaval Hall is to enter on to a wondrous stage set. The façade has been described as an 'astonishment'. It is an unforgettable example of Vanbrugh's originality and masterly disregard for the rules. He imbued all his great buildings with a drama that few have matched. It struck me that if someone 'without thought or lecture' could write fresh and witty plays and design some of the grandest buildings in Britain, perhaps even an educational discard like me could achieve something. And I was off to the shore to get started.

Adjacent to Seaton Sluice is the rocky arc of Collywell Bay. My project was to strip off all the animals and plants from discrete areas of the shore and see what ensued. Would similar creatures recolonise? How long would it take?

It was thought that the order of succession was fixed, because earlier colonists somehow paved the way for later arrivals. That's not what happened on the shore where many of the squatters settled together as soon as the space appeared. The order in which they became apparent depended on their growth rates, not their arrival times.

In the areas I denuded the opportunistic sea lettuces *Ulva* and *Enteromorpha* were the first to show. Soon all the stripped areas were bright emerald squares in a field of khaki wrack. But slowly the ephemeral green algae died down and the larger brown seaweeds regained their lost territory by overgrowing and overshadowing everything else.

In its modest way, it demonstrated to me that communities were not static and immutable; they could be changed at least transiently by a single disturbance. Of course, the initial perturbation need not be deliberate. It might result from the effects of a hard winter or a good year for the recruitment of grazers, or a catastrophe like an oil spill. What is more, shore organisms were the ideal test bed for studying such changes in natural communities because the players in the drama – the snails, limpets and seaweeds – were relatively short-lived and therefore the speed of change and recuperation from disturbance was conveniently rapid.

I had also learned something far more important: that scientists find out by *doing*, not by being told. Existing ideas and facts are merely signposts to the future and sometimes misleading ones at that. To linger too long on what is already known, however interesting that may be, is to be distracted from the business of science, which is not just the accumulation of facts, but the pursuit of the new and, if you are lucky, the unexpected. On this unremarkable shore I had begun a quest that would last a lifetime.

My love affair with the sea would lead me to walk on empty shores before the tourists invaded. And I would come to know the ocean as a wild aquarium, a laboratory, a cemetery for men and ships, and an anthology of legends. I would understand the spell it casts on fishermen, divers, treasure seekers and writers, for the sea is a place for obsessives.

One of my early obsessions was geology so I took a hammer and chipped away at the shale. A school party ambled down on the beach.

'Are there fossils around here?' the teacher enquired.

'A few,' I replied and prised up a sheet of shale. There, pressed as flat as a flower in a book, was the impression of a fossil tree with snakeskin bark.

'Of course, you have to know where to look,' I said modestly.

'Ooh,' she said. 'Do you want that one?'

Well, clearly a hotshot palaeontologist who dug these things out by the dozen could hardly refuse.

I never came across another tree, although down the local pit they often found them and their roots, and even the occasional coelacanth. My finds were more modest – ancient marine mussels huddled together awaiting discovery, and tiny hacksaw blades called graptolites (rock writing), many of them in pairs like long-silent tuning forks. They were an enigma, but are now considered relatives of the sea squirts. In life, some hung from floats as a fringe adorns a Victorian lampshade; others stuck in the mud like fallen darts. They tried every variant to survive, but still they became extinct, destined to sleep for ever in the rocks, in what even geologists call bedding.

These beds had once been tropical swamps. Vegetation was

pressed beneath the crush of fallen forests and accumulating sediment until only carbon remained. We call it coal and burn it to release the energy of sunlight that dappled leaves over 300 million years before.

Northumberland was once riddled with coal mines. The nearest pit to Collywell Bay was at Hartley where, in 1862, the twenty-ton engine beam that pumped out water crashed down the main shaft, filling it with rubble and cutting off the air supply. Two hundred and four miners lost their lives, some of them as young as eleven. Apart from five who were killed outright, they all suffocated, and when the rescuers tunnelled in they found them lying in rows, the boys with their heads on their fathers' shoulders.

Some of the mine's drifts tunnelled out under the sea for over two miles. The seams were rarely even a yard high. Imagine having to crawl on your stomach in the darkness with the ocean rumbling overhead.

One day I noticed an opening halfway up the cliff. Could it be an adit, an old shaft coming out on the cliff?

I climbed up and found that the hole was only 16 or so inches square, but there was an area almost as big as a door that had been filled in and disguised as part of the cliff face. The hole was where some of the fill had fallen out. The next week I returned with a torch, climbed the cliff and squeezed into the hole. I was in a man-made gallery about 170 feet long, which in the gloom seemed never-ending. Suddenly it opened out into a room and beyond this was another and another and another. It was no mine; it was a cave of empty rectangles hidden thirty-seven feet underground. From the ceiling hung long thin stalactites of lime. There was a staircase to the surface, but it was blocked by rubble and soil, the only way out was the way I had come. The quiet was damped down to way below silence and outside the feeble disc of torchlight the darkness was absolute. Then the light yellowed and began to flicker. I returned to the tunnel before I was lost for ever in the dark. If I should become trapped here, nobody would ever come to search. I thought again of the miners tunnelling beneath the sea.

Who had hidden down below in these secret burrows? The answer, I discovered, lay forty-four years in the past. On the morning of 16 December 1914, Winston Churchill leapt from his bath when handed a dispatch that read: *German battlecruisers bombarding Hartlepool.* Well over a thousand shells, some weighing a ton, battered the town, killing 112 civilians and an army private, the first British soldier to die on British soil in the Great War. The ships lobbed another thousand shells into Scarborough, holing the lighthouse and wrecking the Grand Hotel. A dazed man in his nightshirt was seen wandering the streets clutching a Christmas cake. Indeed, the whole of Scarborough was dazed.

A British battle squadron was sent in pursuit, but the three enemy cruisers vanished into the fog. In fact, we had been lying in wait for the Germans. Long afterwards it was revealed that the British had captured a German codebook and knew *when* they were going to strike the east coast, but not *where*.

In the weeks that followed, recruitment rates soared. The Admiralty awakened to the vulnerability of the River Tyne and supplied an immense gun turret from HMS *Illustrious* to perch above a shellproof complex in the cliffs at Collywell Bay. Twenty men sheltered below ground waiting for the enemy to reappear, but he never came. Indeed, the guns were not even test-fired until four years after the armistice, and this caused far more damage to Seaton Sluice than the Kaiser ever did.

Later, the guns were removed and all traces on the surface were obliterated, except for the water tower and a 'defensible latrine'. The Hun wasn't going to catch us with our pants down.

I needed to pass some more exams. For practical tests we had to go to the local university at Newcastle. It would include a dissection and I had examined numerous rats, frogs and dogfish in preparation. Waiting nervously outside the examination room, one candidate peeped through the half-obscured window in the door. He saw a carcass lying on the bench awaiting us. 'It's a squid,' he gasped in disbelief. None of us had even seen a squid before, let alone cut one up. We slumped against the wall and one lad fainted.

Fifteen minutes later, when we were allowed into the room, the kraken (perhaps a technician's idea of a joke) had vanished and we each had our old friend, the dead dogfish.

I needed a bursary if I was to go to university so I didn't just take advanced-level exams, but scholarship level as well. I recall there was a question on island faunas. I knew all about how isolated islands went their own evolutionary way and gave rise to Darwin's finches, and, when American marines fought their way across the Pacific, as John Wayne leapt ashore so did dozens of rats who extinguished the unique island pigeons and the endemic rails. The only problem was that I couldn't remember the name of a single island. So I invented entire archipelagos: Pangoa, Tonalongi, the Malawonga group. I guess no one else could remember them either, for I was awarded a County Major Scholarship.

The Sea of Surprises

Northumberland coast

5/2067007.

I had been an unruly child, but as I learned more about living things I became too busy to be bad. I was preoccupied observing wildlife in the nearest I could find to wilderness.

The northern coast of Northumberland is a wild place of clifftop castles, mist and waves. A huge carcass had washed ashore on a lonely beach. The body was almost thirty feet long and looked like a decaying hippo with fins plus two long stumps – the remains of legs. The neck was now reduced to a badly wrapped, serpentine line of vertebrae with the bristles of the mane clearly visible on the narrow head with its sinister empty eye sockets. Grizzled fishermen had never seen anything like it. It was indisputably a sea monster.

Biologists scampered to examine it and declared it to be a basking shark. The 'legs' were the claspers, long copulatory organs that were thrust into the female when the opportunity arose. The 'mane' was the remnants of the bristles the fish used to filter plankton from the sea for its dinner. But what about the neck? Well, dead basking sharks easily lose their lower jaw, as it is large and poorly attached. The skull is a flimsy affair for its size, for filter feeders have no need of the heavy cranium and massive muscles of predators. Apart from broomloads of bristles, it consists of a mouth the size of double doors and a brain little bigger than a golf ball. Basking sharks are, in all senses, empty-headed. Once the jaw becomes detached, all that remains is the pointed snout on the end of the spine that looks as if it must have once been a long neck. It was indisputably a basking shark.

Everyone was disappointed. They *wanted* it to be a monster. Like John Steinbeck, they felt that an ocean without monsters is like a sleep without dreams. We *relish* the idea of the dark depths being asquirm with forgotten dinosaurs. But we haven't always done so.

For centuries the shore and the ocean beyond were places of terror. Although wealthy Greeks and Romans enjoyed the proximity of tamed segments of the transparent Mediterranean, they hated and feared the ocean, which was populated with gods and their relatives, many of whom were capricious or, like Medusa and Scylla, deadly.

More northern peoples were repulsed by their opaque, icy seas. The ocean was seen as unpredictable and evil-tempered. It roared with anger, and maelstroms swallowed the unwary. There was not even Joseph Addison's ambiguous appreciation of the 'exquisite horror' of the tempest. The coast was a place of drownings and shipwrecks and an open door for the ingress of pirates and invaders. Even infections were believed to originate from the ocean's 'horrid miasmas' that wafted ashore; the idea of fresh air blowing in was inconceivable. The horizon did not beckon, it merely reminded humans of the limitless void. Not until the late 1800s would shore dwellings be commonly built to face the sea.

The Bible had not helped the ocean's image. Eden was a garden far from the sound of the threatening surf, and the deluge came as God's punishment. The shore was therefore not merely the meeting of land and sea, but the battle line. The ragged coasts were the ruins left by the assault of God's wrath against the land. The ocean floor was envisaged as the 'most frightful sight that Nature can offer', as it was 'so vast, so broken and confus'd, so every way deformed and monstrous'.

The deep was also infested with monsters, which Milton imagined huddled together in palaces drowned by the Flood, copulating in bedrooms abandoned by sinful humans. The deep was the place of the Devil and all his devilish creations: the giant octopus, kraken and sea serpent. There are dozens of depictions of these monsters based on eyewitness accounts. Ships are shown enveloped in tentacles, or threatened by dragons whose heads loomed high above the rigging. We were petrified of what James Elroy Flecker called 'the dark . . . serpent-haunted sea'.

The reality of the sea serpent was given credence not only by frequent sightings (over three hundred encounters in the nineteenth century alone and more since) often by experienced and trustworthy observers, but also by the discovery of huge fossil sea creatures such as snake-necked plesiosaurs that were being wrenched out of the Yorkshire cliffs near Whitby. The most widely read Victorian naturalist, Philip Henry Gosse, wrote a detailed study of the sea serpent and was clearly a believer. Most observers described the small head perched on a long neck and many recalled a line of humps following behind. If these were the undulations of a swimming serpent, then it was indeed a unique creature, for all snakes whether land-based or aquatic slither side to side not up and down.

The most photographed and fêted monster is, of course, Nessie. She appeared intermittently on the surface of Loch Ness, which at twenty-four miles long and almost a thousand feet deep is easily large enough to house whole tribes of beasts. Nessie is a creature of whim who rarely surfaces if sought, and exhibits infinite variety: sometimes she comes as a reptilian ostrich, legs submerged and wading in the shallows, or as an upturned boat, or a hosepipe out for a swim or, at her most coy, as the wake of an unseen ship seen by a half-awake observer. In 1933, she even went for a tramp overland and Mr and Mrs Spicer almost wrecked their car when Nessie, her neck undulating like a scenic railway, stumped across the road in front of them.

No less a person than Sir Peter Scott became concerned that Nessie, never having been officially named (and therefore, as far as science was concerned, non-existent), was open to all sorts of exploitation and in need of the protection afforded by the Conservation of Wild Creatures and Wild Plants Act. So, based on underwater photographs computer-enhanced by the Jet Propulsion Laboratory at Pasadena, he described Nessie as fat, seventy feet long with a small head perched on thirteen feet of neck. Another photograph indicated that she had large diamond-shaped paddles, so he named her *Nessiteras rhombopterix* – the diamond-finned marvel from Ness.

His description was published in the most prestigious scientific journal in the world and perhaps it is only by chance that an anagram of the name of Scott's co-author, Robert Rines, is 'Bets in error'. But is it also a coincidence that *Nessiteras rhombopterix* can be rearranged to spell 'Monster hoax by Sir Peter S'?

Even the most seductive creatures could be lethal. Maritime folklore is full of tales of mermaids singing to lonely sailors, luring them into danger as surely as a wrecker's lantern. They perched seductively on rocks admiring themselves in a mirror and combing their long hair. Flowing locks seem unlikely after spending every day in salt water before the advent of Wash and Go.

Nowadays mermaids seem to be extinct, perhaps a victim of pollution or the invention of the camera, but they were a danger in their day. According to Columbus they were not as beautiful as depicted. But their sad song was irresistible and they posed riddles that seamen befuddled with a glimpse of breasts could never hope to solve. They dragged mariners down to their level – below the waves.

They were both desired and feared. Odysseus wisely blocked his crew's ears against the siren's song. Caligula's troops mutinied rather than cross the mermaid-ridden English Channel. The invasion of Britain was almost called off.

In Tudor times 'mermaid' was synonymous with harlot. However seductive the top half of a mermaid might be, the prospect of fornicating with the scaly underpart has limited appeal. Magritte's image of a 'Practical man's mermaid' has the head of a fish, but the lower reaches of a woman.

The rational explanation for mermaids seems even less plausible than their existence. It is claimed that manatees (in the Atlantic) and dugongs (in the Indo-Pacific) might easily be mistaken for foxy or, rather, fishy maidens. They officially belong to the Sirenia, the sirens.

The name 'manatee' conjures up the scent of an Indian princess, but dugong better captures the spirit of the beast. They resemble seagoing settees and have outstanding halitosis. I would have to have been at sea for a very, very long time before I would be likely to see a dugong, sniff its breath and then leap over the side to my doom.

Just occasionally, live fish women washed ashore. In 1403, one covered in 'sea mosse' fetched up in Holland and lived for fifteen years in Haarlem without being able to master a word of the language and was forever trying to return to the sea. She was taught to kneel before the cross and forty clergymen vouched for her authenticity. A winged mermaid was caught off Exeter in 1730, and a particularly large-breasted specimen washed up in the Hebrides and was given a Christian burial.

The legend lived on in fairground freak shows. The Victorian naturalist Frank Buckland examined a yard-tall exhibit, but found only a papier mâché torso with the hindquarters of a haddock and fish's teeth for dentures. In Egypt, there was a brisk trade in specimens that were the top half of a monkey perched appropriately on a Nile perch. They often came with an affidavit of authenticity. The latest sighting of a mermaid that I recall was reported in a Philippine newspaper in 1994 under the heading MERMAID DROWNS TWO YOUTHS.

The manatee has been designated a threatened species. We have starved them by converting their meadows of sea grass into prawn farms, entangling them in nets and slicing them with the propellers of passing boats. Perhaps these doleful innocents will soon be as extinct as the mermaids.

Some of the most monstrous monsters proved to be real. One was described by the Bishop of Bergen in the mid eighteenth century:

> It is called kraken . . . which is round, flat, and full
> of arms . . . Fishermen unanimously affirm . . . they
> see this enormous monster come up to the surface . . .
> though his whole body does not appear . . . its back,
> which seems in appearance to be about an English
> mile and a half in circumference (some say more,
> but I chuse the least for greater certainty) . . . horns
> appear, and sometimes stand up as high as the masts
> . . . It seems these are the creature's arms and, it is
> said, if they were to lay hold of the largest man-of-
> war, they would pull it down to the bottom.

In 1673, a smaller creature fitting this description drifted ashore in Kerry. It had eyes as big as a man's head, a body bigger than a horse and was furnished with tentacles over eleven feet long, the shorter ones covered with suckers as big as crown coins. In 1790, another washed up with eighteen-foot tentacles and an even longer body, so thick that a grown man could just span it with his arms. Not as enormous as the Bergen beast, but nonetheless a monster.

As more specimens appeared, a new species was established, *Architeuthis monachus*, the giant squid. The kraken was now authenticated and larger ones would come to light, but very few would be caught alive and none have yet been seen underwater, for with the biggest eye in the animal kingdom and the ability to jet around at ten feet per second, giant squid can easily evade anything that approaches and arouses its suspicion.

There is also evidence that they are agile and voracious predators and have heavyweight adversaries. A mariner witnessed a battle from a whaler: 'A very large sperm whale was locked in deadly conflict with a cuttlefish or squid, almost as large as itself, whose interminable tentacles seemed to enlace the whole of its great body.' Who would eventually consume whom was still in doubt, but my money's on the whale.

Such encounters are not unusual. In 1938 a scientist with the Ministry of Agriculture and Fisheries examined the carcasses on a whaling ship and reported that 'nearly all the sperm whales carry scars caused by the suckers . . . of large squids, scars up to ten centimetres in diameter being common'. Suckers of this size indicate an animal with a body perhaps thirty feet long and arms almost twice that. One species of shark feeds almost entirely on *Mesonychoteuthis* (colossal squid). The dimensions of a recently caught juvenile colossal squid indicate that a full-grown adult probably reaches fifty feet in length. Clearly the full-grown kraken still eludes us.

The oceans are vast so there is ample room for even the largest creatures to swim around undetected. It is estimated that there may be millions of giant squid in the sea, but how rarely we see one. It would not be until 1976 that we would discover a thirty-seven-foot, three-quarter-ton basking shark called 'megamouth', or more prosaically in Latin, *Megachasma*, the great yawner.

Surely with ever larger fishing nets, and scientists plumbing deeper and more frequently into the sea, only insignificant animals have escaped our gaze? Can there possibly be monsters yet to be found?

Recently, a systematic survey of the first description of all sea creatures over two metres long revealed that since 1758 we have discovered on average a new species almost every year. The rate is slowing down, but even so it is predicted that we can continue to expect an unknown big animal to be found every five years or so. It may be a new species of crocodile, or squid, or a giant jellyfish, but it *will* be big.

As recently as 1995 a large sea serpent was described for the first time, from a specimen found in the stomach of a sperm whale, and in 1998 a previously unknown species of whale was discovered.

Somewhere, the sea still holds a giant surprise or two.

Sentiment
and Seaweed

Ilfracombe, Devon

That summer, before I left home for good, was the last time I would go on holiday with my parents. We toured Devon in a fat, black Austin that had once been a police car. It still retained the smell of criminals, according to my mother, who had the olfactory apparatus of a bloodhound.

We camped on the clifftops where fulmars circled on the rising air. I watched rabbits munching flowers from the thrift and was fascinated to see dolphins herd mackerel into a small bay then wait for the receding tide to channel them out into their jaws.

I gazed into the tide pools and went snorkelling whenever I could – at Babbacombe Bay and Hope's Nose. A hundred years earlier they were also some of the favourite collecting sites of Philip Henry Gosse.

Gosse was the son of a painter, and was a self-taught naturalist. At the age of seventeen he was sent to Newfoundland to become a clerk to a whaling company. Eleven years later he returned with the manuscript of a book, *The Canadian Naturalist,* in his pocket. In the words of his son, Edmund, the book contained 'the germs of all that made Gosse for a generation one of the most popular and useful writers of his time . . . the picturesque enthusiasm, the scrupulous attention to truth in detail, the quick eye and the responsive brain, the happy gift in direct description'. All he lacked was a sense of humour.

He scraped a living writing books on natural history until a bout of nervous dyspepsia forced him to move from London to Torquay, on the Devon coast. From then on his health improved and the seashore became his passion. Once he had explored the nearby shores, he organised excursions to all over the county and then to south Wales.

Gosse was a devout member of the 'Bretheren' and his son endured a stern childhood. His father even rejected Christmas pudding as 'the devil's sweetmeat'. But as Edmund grew he accompanied his father on 'laborious and exquisite journeys down to the sea . . . laden with implements . . . it was in such circumstances as these that my Father became most easy, most happy, most human'.

Charles Kingsley became a family friend and without Gosse's influence *The Water Babies* would never have been written. Gosse was sometimes a trying acquaintance. When Kingsley came to call, Gosse instructed the maid to 'Tell Mr Kingsley that I am engaged in examining the scripture with certain of the Lord's children', and the guest, his hawk's-beak face glowering, paced about the garden waiting to be received. Even so, Gosse so imbued Kingsley with a love of marine biology that it was he that often dragged them out to go dredging in Torbay. Young Edmund found his jolly exuberance a welcome antidote to his father's unremitting seriousness.

In 1853, Gosse published *A Naturalist's Rambles on the Devonshire Coast*, the first of a series of chatty but learned books on seashore life accompanied by his own charming paintings. They were just what the public craved. They satisfied the Victorian yearning to be 'improved'. Even more, they provided folk with something to do on their annual summer migration to the coast to escape the city smoke. In every remote creek and fishing village, families of town dwellers came to dig the sand or turn the boulders to discover treasures abandoned by the tide. A guide to Scarborough extolled the 'variety of seaweed and coralleries, pebbles and petrifactions to be met with on the rocks'. Books advertised as suitable for 'railway and seaside reading' included such page-turners as *Seaweeds: Instructions to find, preserve and classify them*. For the middle classes, no seaside sojourn would have been complete without excited talk of zoophytes and polyps, feather stars and fucus.

Victorians were stirred by those 'exciting emotions which the view of Nature, in its bolder phases, produces'. In a contemporary engraving, a well-to-do couple in their Sunday best are promenading along the strand. She clings to his arm and he clutches a kelp plant. The picture is entitled *Sentiment and Seaweed*.

Gosse became a household name and his books spawned many imitators; even Charles Kingsley wrote a best-selling guide to marine biology, as did George Eliot's common-law husband, George Henry Lewes. The latter described the gear required by collectors: 'an old coat, with manifold pockets in unexpected places . . . Trousers warranted not to spoil; over the trousers are drawn huge worsted stockings, over which again are drawn huge leathern boots.' Gosse ignored all such advice. Even on the shore he dressed like a priest and was seen 'plunging into a pool in full sacerdotal black, after a sea anemone'.

Ladies were warned that to explore the shore they 'must lay aside all thought of conventional appearances, and be content to support the weight of a pair of boy's shooting boots . . . rendered as waterproof as possible by a coat of Neat's foot oil . . . Next to boots comes the question of petticoats . . . to make the best of a bad matter, let wool be in the ascendent . . . and never come below the ankle . . . A hat is preferable to a bonnet, merino stockings to cotton ones, and a strong pair of gloves is indispensable. A stick is a very desirable appendage, for drawing floating seaweeds from the water, but . . . people amuse themselves by devising ingenious varieties. The basket may be exchanged, by those who care to invest in it, for an Indian-rubber bag, which can be strapped round the waist . . . Feel all the comfort of walking steadily forward confident in yourself, and let me add, in your dress.'

Gosse led excursions of amateur enthusiasts to the shores near Ilfracombe where 'the ragged rock pools that lie in the deep shadow of the precipice [are] tenanted with many fine kinds of algae, zoophytes, crustacea, and medusae'. Edmund described one of these trips: 'At the head of a procession, like Apollo conducting the Muses, my father strides ahead in an immense wide-awake [hat], loose black coat and trousers, and fisherman's boots, with a collecting basket in one hand, a staff or prod in the other.' He recalls his godlike father standing against the dark rock at the mouth of a funnel in the rocks through which a roaring jet of foam sporadically exploded and encircled him in a rainbow. On another occasion he was seen lecturing to women who lay 'spread-eagled on the sand' scratching for shells, with their wool in the ascendant.

Gosse coined the word 'aquarium' and taught keen collectors how to keep their treasures alive. He also devised a formula for artificial sea water to avoid 'the inconvenience, delay and expense attendant upon the procurement of sea water from the coast'. Marine aquaria became the latest fad and in 1853 Gosse stocked the first ever public aquarium, at the Zoological Society of London. Back on the shore, the emphasis was now on collecting, not looking, and collecting was not a gentle art. Lewes recommended that naturalists should be armed with a 'hammer, chisel, oyster-knife, and paper-knife'. The collecting cry of the naturalist Thomas Rymer Jones was 'Come, boy! The fishing basket and the muslin landing net — a hammer and an iron chisel.' Kingsley preferred 'a strong-backed quarryman, with a strong-backed crowbar.' The effect on shore communities was devastating.

In later life, Edmund Gosse reflected sadly on the pristine pools he had seen fifty years before:

> If anyone goes down to those shores now . . . let him realise at once, before he takes the trouble to roll up his sleeves, that his zeal will end in labour lost. There is nothing now where in our day there was so much . . .
>
> All this is long over, and done with. The ring of living beauty drawn about our shores was a very thin and fragile one. It had existed all those centuries solely in consequence of the indifference, the blissful ignorance of man. These rock-basins, fringed with corallines, filled with still water almost as pellucid as the upper air itself, thronged with beautiful sensitive forms of life – they exist no longer, they are all profaned, and emptied, and vulgarised. An army of 'collectors' has passed over them, and ravaged every corner . . . That my Father, himself so reverent, so conservative, had by the popularity of his books acquired the direct responsibility for a calamity that he had never anticipated . . . cost him great chagrin. No one will see again on the shore of England what I saw in my early childhood, the submarine vision of dark rocks, speckled and starred with an infinite variety of colour, and streamed over by silken flags of royal crimson and purple.

The Victorian craze for collecting seashore creatures eventually subsided, but with increasing pollution and coastal development some shores were never quite the same again.

And visitors remain a problem. A survey of Devon shores a few years ago revealed that the construction of a car park adjacent to the coast was the second-worst thing that could happen to intertidal communities. To be situated near to a marine laboratory was worst of all. At one laboratory the staff confessed that when they returned from serving in the army during the Second World War, they were amazed how rich the shores had become in their absence.

Fewer biologists now lead marine excursions to Ilfracombe. But nearby, on a slaty shore beneath the tall, hog's-back cliffs, I found a fish high and dry on the rocks. It was a lumpsucker, a swollen, humpbacked unfish-like fish, with tubercles on its skin instead of scales. It hadn't flopped over in that fishy way; the sucker beneath its chest clung to the rock and kept it upright. It had ventured over the shore at high tide and been stranded by the retreating water. Sometimes after storms they are cast ashore in their hundreds. This one was nursing a flicker of life until the tide returned, but would probably succumb to drought or be disembowelled by crows before then. It had large round eyes that seemed to be wondering how it had come to be there in the first place.

I am holding a photograph of Philip Henry Gosse. He has that same plaintive look.

Navel
Manoeuvres

Liverpool

In 1960, having left home on my twentieth birthday, disguised as an adult, I went off to study biology at the University of Liverpool and moved into lodgings.

It was one of those severe houses that had once been the home of a respectable merchant. He wouldn't have recognised it now. In the estate agent's jargon, it needed a little work. The rendering was loose and the window frames were a battleground where rival fungi fought to consume the remaining wood. More flowers blossomed in the gutters than the garden.

I heaved on the doorbell. Mrs B answered the door and I handed her the bell pull.

'It just came away in my hand.'

'Not to worry, everything's always coming adrift,' she said with the echo of an Irish brogue, stuffing the bell pull back into its hole to surprise the next caller.

Mrs B had an explosion of red hair, arms splashed with freckles, and a smile to warm your hands by. She wore a wrap-around pinny and, like my granny, used it to hide acres of unlikely anatomy.

'You'd better come in before you drown.'

I followed her upstairs, averting my eyes from her calves of palest corduroy.

'And this,' she said with misplaced pride, 'is your room. Come down for a nice cup of tea, then you can unpack and get acclimatised.'

It would not be easy.

The bed was of tarnished brass rails and thistle-topped knobs. I sat on the counterpane and surveyed the room as best I could. The glimmer from a dusty sixty-watt bulb gave up before it reached the corners. Grey net curtains clung to the grey wet windows. No light would enter there.

Everything was huge. A slab of veined marble with a decorative fat-and-blood design stood on iron legs with great claw feet. The wardrobe loomed over me, its long mirror cracked from top to bottom so that my reflection seemed to be trying to flee in two directions. But there was no escape. For my first year at university this would be home.

In the Swinging Sixties temptation was for me the landlady's daughter intent on navel manoeuvres. Mrs B's Doris liked to chat to me when I was on my own. 'My friends are very bohemian,' she boasted. 'We smoke cheroots and sit on the floor listening to Ravi Shankar.'

'Sitting on the floor?' I said. 'Gosh!'

Matters took a more sinister turn at six o'clock one morning when I was awakened by Doris wearing a filmy nightie.

'I thought you might like a little something special.' She was offering it to me on a plate. It was bacon sandwiches.

Mrs B had been a Roman Catholic and Mr B a Plymouth Brother so both their families wept when they married and fainted when they became Mormons. Waves of evangelists were dispatched from Salt Lake City to the furthest corners of heathen England. As Mr B was now a lay bishop in the church, it was his duty to host them for free.

There was much preparation for the arrival of the guests. Even the tablecloth might be degreased or the three-piece suite shampooed. This was best done in the garden, so the dark grey settee and armchairs were carried out to sit on the grass like stranded dugongs. Within minutes they were covered in sudsy froth and revealed to be beige. I began to feel queasy.

The guests arrived a day early and helped us to carry the sodden furniture indoors.

'Sit right down and I'll be putting the kettle on for some *Instant Postam*,' said Mrs B with an air of quiet excitement, as if she'd offered them a sniff of cocaine.

They sat down with an audible squelch and little iridescent bubbles popped from the sofa and drifted aimlessly away. Every time they adjusted their damp bottoms more bubbles emerged. I stood casually and tried to conduct a normal conversation.

They were crew-cut clones with their patter off pat. 'Hi there! I'm Randy. It's truly wonderful to make your acquaintance.'

'How will you spend your time?' I asked.

'In cemeteries mostly, I guess. Baptising the dead.'

'Won't the living let you in?'

Doris never offered me bacon sandwiches again.

Her mother noticed a change in her. 'Doris looks very pale. Do you think she's coming down with something?'

I said nothing.

One evening I was invited into the parlour for a cup of tea and a chat with Mr and Mrs B. Mrs B had excavated some ancient biscuits from the back of a cupboard. When she proffered them I thought she was going to ask if I could get them carbon dated at the university.

Mostly we talked about evolution. They were both creationists and I felt obliged to champion rationality.

'How then do you explain the fossil record?'

'Quite simply. God put fossils there to test our faith. To show us that we must believe *in spite* of the apparent evidence.'

'But they're arranged in the strata in order, going from simple forms to complex.'

'Well, of course. Otherwise it wouldn't be a test, would it? God knows we're only misled, not stupid.'

'But we can see natural selection in action. Take melanism in moths, for instance. I'll try to explain it simply. There are these moths –'

'Peppered moths you mean, *Biston betularia*?'

Oh dear, they knew more about this than I did.

Their arguments had all been rehearsed over a hundred years before by Philip Henry Gosse. When he had gazed into those Devonian rock pools, he had seen not just the wonders of nature, but also the workings of the Almighty. 'The gratification of scientific curiosity,' he argued, 'is worse than useless if we ignore God.' Although he admired Charles Darwin, he was appalled at the implications of the theory of natural selection and was determined to champion the creationist cause.

The nub of his argument was that on the day of creation God made everything exactly as we now know it. So trees had annual rings in their trunks, petrified palms had the scars of last year's lost leaf stalks, rocks had fossils embedded within them. Everything had the signs of a prior existence, a fictitious past. He claimed that these manifestations reflected not the passage of real time, but of *ideal* time in the mind of God. Gosse believed that one 'cannot avoid the conclusion that each organism was from the first marked with the records of a previous being'. He asserted 'not that it could have been thus, but that it could not have been otherwise'.

His case appeared in a book called *Omphalos*, the navel, referring of course to Adam's navel. This had once been a hot topic of debate, a matter that no fig leaf was big enough to hide. If Adam lacked a navel then he was not a perfect human being and surely God would not create an imperfection, but since Adam had not come from a placenta why would he require a navel? Would God fabricate such an object for no purpose? Gosse had the answer. He published and sat back waiting for his thesis to rally all the opposing forces.

His son Edmund, having escaped the shackles of his upbringing to become one of the leading intellectuals of apostasy, wrote: 'Never was a book cast upon the waters with greater anticipation of success than this curious, this obstinate, this fanatical volume. My Father lived in a fever of suspense.' His book would, he thought, quell all the turmoil of scientific speculation, but that was not the way it would be. Neither atheists nor Christians embraced it. The reviews were scornful and private letters were 'few and chilly'. Even his friend Charles Kingsley, a 'muscular' Christian who had once 'taken on' Cardinal Newman, wrote to say that he could 'not give up the slow and painful conclusion of five and twenty years study of geology, and believe that God had written on the rocks one enormous and superfluous lie'. As more and more letters like this arrived at the Gosse household, gloom descended upon their breakfast table.

Edmund was hard on his father, perhaps because he was embarrassed and disappointed that Gosse had been so close to the heart of the great revolution, but chosen the wrong side.

Gosse never recovered from the shock of having offended almost everyone, perhaps even God, in his attempt at universal reconciliation. He still believed that he alone had uncovered the truth. Why then had God abandoned him? For what sin was he being punished?

Gosse finished his delightful monograph on the sea anemones, but then retired from science. If only he had been born in quieter times his reputation would have survived unblemished. Instead, he was seen as a man clinging to the verities of the past just as science changed direction towards the modern age. Darwin's obituary in the

Proceedings of the Royal Society ran for thirty-eight pages, Gosse was allocated three.

Forty years later I would come across a newly published creationist booklet entitled *Did Adam Have a Belly Button?*

Evolution is a very slow process.

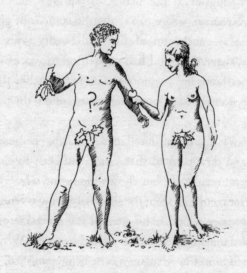

Bunnies and Badgers

Anglesey, North Wales

My teachers at Liverpool University were a touch eccentric. One used to lecture with his shoes off and his feet lodged in a desk drawer. He balanced way back on his chair reading out his notes as we underlined the identical passages in the recommended textbook. Another hid one day behind a free-standing blackboard prior to class and, after we had all trouped in and were chatting, he sprang out and announced, 'If you are not going to listen, then I'm off!' and stormed out.

The old professor was an authority on limpets and decided to lecture on his favourite topic dressed as a limpet. He appeared with a large cardboard shell on his back, antennae dangling from his head and various organs pinned to his chest. I suppose he intended to point to his appendages and uncover hidden organs as appropriate during the talk. But he never got to open his mouth for the students laughed and jeered and stamped. He exited, never to lecture again.

University was a crowded curriculum of girls and parties and it was only a conscientious objector to these that had any time to study. I tackled chemistry with the confidence that comes from complete ignorance. Yet I understood it, or so I thought until the exams proved otherwise.

There were two lecturers on the same botany course: Bunny

Burrows was bustling and bristly, Peter Dixon bluff and assertive. They were as different as chert and chisel. Sparks were certain to fly.

I remember Dixon's first lecture.

'The only text on the red algae is ten years out of date,' he began. A clutch of tardy students came in and clattered to their seats.

'As I was saying, the only book on the red algae is –'

More latecomers, more clatter.

'The bloody thing is almost eleven years out of date!'

Bunny Burrows was the only female member of teaching staff in the department. Science was (and amazingly still is) largely a male preserve. A hundred years ago women were banned from the biology sessions of the British Association for the Advancement of Science because of 'the nature of the papers to be presented'. What was it that might have shaken women to their foundations? Mr Allis giving a talk on 'The toes of an African ostrich'? 'Vegetable monstrosities' perhaps, or 'A fish with four eyes'. No, I think it was Mr Wallace's account of 'Inosculation observed in two trees' – which proved too risqué for even an abstract to be published. In 1902, a woman was put up for a fellowship of the Royal Society, but rejected on the grounds that she was ineligible because as a woman she wasn't a person. This behaviour was not confined to Britain. Marie Curie, who was awarded *two* Nobel prizes (and in different subjects), was rejected for membership of the Academie de Science, the French equivalent of the Royal Society. Even today only 4 per cent of the Fellows of the Royal Society are female.

Bunny felt that to be as good as a man she would have to be better. As Ginger Rogers once said, 'I did exactly the same as Fred, but had to do it backwards and in high heels.' Bunny was hypersensitive, so that if she came to coffee late and the jug was empty she assumed that the men had slurped extra cups just to deprive her.

She always kept a motherly eye on the female students and a stern watch on the males. I guess I was her most troublesome male.

I will never forget that fearsome look when, as usual, I had incurred her wrath. She hated having her photograph taken and I took this as a challenge.

'You took my photo,' she bristled after hearing the click of the shutter.

'No,' I replied, 'I just glanced in the viewfinder, focused and pressed the shutter like this.' Click!

'You did it again!'

We had a mutual admiration/exasperation relationship. Even so, when she took students away on field trip I was enlisted as a demonstrator. One evening, just as the students were gathering to go to the pub, Bunny took the demonstrators aside and asked: 'Would you like to go and see the badgers?'

'Badgers?' I repeated in disbelief. 'Why yes, of course. Badgers.'

We sat on a windswept hill trying to make out the set in the darkness. And we sat and we sat and we sat. The badgers, assuming they lived there at all, had trundled off at dusk and wouldn't be back until dawn. I was sinking into an abyss of boredom.

'Should we retreat to the pub?' I said and the terrible words echoed from hill to hill.

'Yes, let's,' said Bunny, who was probably as anxious to leave as I was, but couldn't admit it. We piled into her car and drove right past the nearest pub. We didn't stop until we reached a quiet hostelry three villages away. I guess no student was likely to see us tippling there and she could relax and enjoy herself.

The next night she came up to me and said with a wink, 'Shall we go to watch the badgers again?' Every night for the rest of the trip we watched the badgers until we were experts on the subject.

The nearest decent rocky shores to Liverpool were in North Wales, where caravans went to breed and you needed a pint of phlegm to pronounce the place names. Few students had cars in those days so Bunny rented one that was two sizes too big for her and peered nervously at the road through the steering wheel. Rod, one of the students, drove her new Triumph.

We headed for Rhosneigr, a village of neat little streets and whitewashed houses on Anglesey. This was the most fiercely nationalistic corner of Wales. We visited a local pub for lunch and the moment we entered, the locals switched from English to Welsh. On the jukebox Tom Jones sang 'It's not unusual'.

We headed for two lagoons separated by a low ridge of rock. 'That is where we will collect,' said Bunny pointing to the far lagoon.

'Last time we collected in the nearer one,' Dixon argued.

'No we didn't. It was that one over there.'

'I distinctly remember . . .'

This went on for some time, getting more heated all the time. Meanwhile, the tide had turned and flooded the ridge. The two lagoons became one.

I went snorkelling in the shallows. The water was surprisingly warm in November. The sea lags behind the land and is coldest in the spring and warmest in autumn. Water is adept at absorbing heat without warming up, that's why we use it to extinguish fires. It takes a whole summer's sun to warm the ocean.

I swam through long tangles of bootlace weed, some coiled like bed springs. Gliding over the sand were thousands of brownish-maroon sea hares, three inches long, enormous slugs with waving frills on their backs and long palps on their heads like rabbits' ears. They look like hunched bunnies made from an ingeniously folded hand-kerchief. My girlfriend's grandfather used to make just such a creature that would nestle in his palm then suddenly shoot up his arm to giggles of delight. I hope that such magical skills have been passed on to the present generation of grandfathers.

Sea hares are creatures halfway between snails and sea slugs. Perhaps they were snails long ago, but evolution determined that the design could be improved. A snail carries a coiled shell on its back and, to fit inside, folds its body so that only the most important bits protrude – its head and anus. Unfortunately, the anus is sometimes situated *above* the mouth. Some groups gradually did away with their shell and the sea hare is almost there. Invisible inside its body is a calcareous memento of a shell.

Lacking defensive ramparts, the sea hare squirts out purple ink to confuse and irritate its predators, indelibly staining the hands of marine biologists. If confined in a bucket, they can suffocate in a sepia cloud, condemned by their own ink – a warning to all writers. In ancient Rome it was an essential ingredient of potions that would put unwanted relatives out of your misery.

For some reason these gentle browsers had acquired an unjustified reputation as one of the most abhorrent creatures in the sea. In 1555, the Scandinavian cleric who had described the kraken also mentioned that although the perfidious sea hare 'doth cause terror in the sea: on land he is as the poor little hare, fearful and atrembling'.

The hermaphrodite hares come inshore to chomp the sea lettuce and lay skeins of orange eggs over the weed. They never return to the ocean, for after spawning they die. Like bunnies, sea hares have an active sex life. A marine biologist who observed them for nine hours found that they copulated for 65 per cent of the time. A single sea hare can lay 470 million eggs in its lifetime. Such sexual exuberance is commonplace in the sea. The most fecund crabs produce two million eggs a year, a starfish extrudes two and a half million at a single spawning, the turbot 10 million, ling cod up to 30 million. These numbers are a measure of their expected mortality. For the oyster, only one egg in 10 million survives to become a breeding adult. No matter how abundant a species is, its future is always uncertain.

Exuberant reproduction is merely a reflection of what a dangerous place the sea is. Few survive the concerted attacks of predators and therefore progeny must be produced in immense numbers in the hope that a few will make it to the next generation. But the failures are *not* wasted. They sustain vast armies of predators, scavengers and decomposers. Death is the engine of life.

The engine almost killed us too. After a long day we set off back to Liverpool. Ahead of us, a truck loaded with steel reinforcing rods pulled over to the left and slowed. As we moved out to overtake, it swung across the road and we slammed into its rear. I dived for the floor as steel skewers came through the window in an explosion of glass. I thought we were dead.

Miraculously, we were almost unscathed. But Bunny's Triumph looked like Bonnie and Clyde's bullet-riddled Buick. The windscreen had gone, the bonnet was concertinaed and the saloon a colander. To our amazement, when we extricated it from beneath the truck and pulled out the wings, the wheels still turned and the engine worked. So we crawled back home to Liverpool. It would be all right so long as it didn't rain.

It poured.

Filtersville

Menai Straits, North Wales

For me, diving became an integral part of studying biology and whenever I could, I returned to the shores of Wales.

Cambria was Roman Wales and the *Ordovices* were the last of the local tribes to submit to Roman rule. The centurions and tribesmen are long gone, but their names linger in the rocks; the Cambrian and Ordovician slates don't just litter the landscape, they *are* the landscape.

Huge chunks have been bitten away and capacious quarries now stand where once there were only hills. Even some of the hills are not what they seem. At Dinorwic the mountain has been hollowed out. Inside is what looks like a set from the finale of a James Bond film, a cavern nine storeys high where enormous turbines spend most of their lives at ease. A languid technician leafs through the *TV Times* and, when *Coronation Street* ends, he flicks a switch so that water falls from the mountain loch above to the lake below and generates just enough power to supply the millions of kettles that are about to be switched on. For the rest of the time water is gently pumped back up to await the finish of *EastEnders*.

A great fault cuts off the square swatch of Anglesey from the rest of Wales. The fault is quiet now, but 500 million years ago, when this was volcano country, it slid and jolted and sent the trilobites scampering for cover. Now it is inundated by the sea and called the Menai Straits.

Fifty miles to the east is the mouth of the River Mersey. The river scours the hills to bring down tons of silt and dispatches them to sea. A ribbon of mud drifts out past North Wales and some is sucked into the tidal stream of the Menai Straits. The six-knot current keeps the mud from settling and in this trough of turbid water the visibility below can be zero.

I dived around slack water at the southern end, beneath Telford's wonderful bridge. About twenty-five feet down the sunlight had been swallowed up and daylight became dusk. Without light, the bottom was devoid of plants. Where the smooth rock sloped away through the murk I saw fields of hydroids and sea squirts, crumbs of bread sponges and sea anemones as big as saucers. All the animals there had but a single purpose – to snatch what they could from the passing flow. Water was there to be sucked up and sieved, for food was floating on the watery wind. They have it easy; all they have to do is to lie in beds all day and drink as much as they like. On land only spiders' webs filter the passing air, whereas armies of animals sieve the particle-rich sea.

Hydroids are tiny tubular trees with a tentacled polyp at each tip, open and anxious to grab anything that moves. The sea squirts, like two-spouted teapots, suck and blow to siphon the sea. That is almost all they can do, for they discard their rudimentary 'brainlet' after a fortnight and so are destined to remain absent-minded and obscure. But it is a case of the children being smarter than the parents, for they have larvae like tadpoles for swimming to the next generation, and it is possible that long ago the tadpole of one of their ancestors went its own way without developing into the adult. It had something the adult didn't, a strand of nervous tissue down its back, perhaps only a few million years away from evolving into a spinal cord, the vertebrates and us.

A sponge is the least pretentious of beasts. It never seeks to impress. It is just a cushion of catacombs lined with microscopic whips to keep its internal currents on the move. The sponge's interior is as labyrinthine as a grand hotel, with all its corridors crowded with tiny sheltering crustaceans that filch morsels from the flow. Outside, the hummocks of sponges seethe with tribes of tiny ghost shrimps that stood up on their hindmost legs and grabbed at passing particles with their claws. The bigger males jostled the females aside to get the best bits, although, by doing so, more of them ended up in the stomachs of fish.

The most abundant filterers of all coat the adjacent rocky shore like a layer of soiled snow. Barnacles are tiny crustaceans, members of the class Cirripedia – 'the hairy feet'. Some, in the best traditions of alien invaders, transform into fungal-like filaments and ramify throughout the bodies of their cousins, the crabs, castrating them and making them forget how to shed their shell; others surf the ocean on the fins of flying fish, but most of them have settled down on the shore and live in calcareous tepees. The trapdoors in the roof can be closed to keep the moisture in and opened to allow for feeding. The creature inside stands on its head and pulls food into its mouth with its legs. They splay beautifully like the ribs of a lady's fan and flutter with a rhythmic grace at up to one beat every second. They comb the currents for food, and the legs, which are as hairy as a chimpanzee's, have bristles that interlock into the finest mesh so that little escapes. If you hammered a speck of dust into a dozen pieces, they could still be caught in the barnacle's seine net.

Barnacles are anchored to the rock by glue that sets rock hard and is impervious to acids, alkalis and extremes of temperature – so good that it was examined as a possible fixative for tooth fillings. It is twice as strong as the glue that holds spacecraft panels together.

If you spend your life stuck immovably on waveswept rock, it is important to ensure you get a good site. Barnacles choose with care. Their tiny swimming larvae search for crevices and dimples in the rock, or adult barnacles, or even where recently departed barnacles have left behind a chemical reminder of their stay. When the larva finds what might be the right spot, it tentatively attaches at one end and pivots like a windscreen wiper. If it touches another barnacle, it lifts off and searches elsewhere. Later, as desperation sets in, it becomes less fussy. I had a T-shirt that bore a graph demonstrating this diminishing discretion. It read AT THIS STAGE I'LL SETTLE FOR ANYTHING.

Although seeking personal space for feeding, it needs to be close to other barnacles when reproduction time comes around. But the bold, bisexual barnacle has a prick up its sleeve; its enormous penis is three or four times taller than the shell. Out it leaps, thin and arching and dips into an adjacent barnacle as neat as a nib into a surprised inkwell. The recipient never sees it coming. Sadly, as one barnacle loses its virginity the other loses its apparatus, for the penis drops off.

Underwater, the current grew and I was dragged out into clearer water beyond the shadow of the bridge. In the sunlight the kelps stretched up their brown arms into the current and swayed. They were only waving, not drowning.

Committed
to Biology

Puffin Island, North Wales

From the train on my journey home I saw a knoll of grey limestone about a mile offshore. On the map it looks like a great whale leaving the other Wales behind. I never got to visit the island, which is a pity because it played a small part in the history of marine biology. Its name is Puffin Island.

Puffins are the most gaudy of British birds, with orange legs, a clown's face and an enormous rainbow nose. They hide away in underground burrows like embarrassed guests who were *sure* it was a fancy dress party.

On this small windswept island everything burrowed: the puffins, the rabbits, and the rats that had scampered ashore when a Lithuanian ship went aground on the rocks. There were tens of thousands of puffins then, but a single pair of rats can generate a thousand offspring per year. Great black-backed gulls took some of them, but it was never enough, so the rats prospered at the expense of the birds. The island was once a burial ground and tunnelling rabbits exhumed human bones and left them to litter the ground. This unpromising place was the site of the first marine laboratory on the west coast of Britain.

William Herdman was only twenty-three years old when he was appointed Professor of Natural History at University College,

Liverpool, a chair he would occupy for forty years. In 1885, he mobilised fellow naturalists to form the Liverpool Marine Biological Committee dedicated to the study of marine life in the Irish Sea.

They embarked on research cruises by hitching lifts on local ships. The *Hyaena* passed its later years with the Liverpool Salvage Association, but it was well travelled. After a sojourn in the Crimea it became one of General Gordon's gunboats in China where he squeezed the opposition into the cities and then pounded them into submission with the heavy guns of his riverboats.

The *Hyaena* was flat-bottomed and shallow-draughted, which meant it rocked violently when out at sea, and sometimes while still in harbour. It could only manage three knots flat out. Into a strong current it had been known to progress full steam ahead – backwards. Its main virtue was a heavily armoured hull. When one of the naturalists noticed they were heading for the rocks, he alerted the captain who, with a casual wave of the hand, replied, 'So much the worse for the rocks.'

The naturalists cast the lead to sound the depth, strained out the tiny plankton with fine silk nets, and trawled the sea floor. But, as guests on a work boat, they were frequently frustrated by being unable to stop at the most promising sites or have sufficient time to deploy their sampling gear. After their longest voyage, Herdman admitted that 'Fortunately one of the crewmen died'. It was 'fortunate' because the boat anchored for the funeral. While the sailor was being committed to the deep on one side of the ship, the biologists hauled up specimens on the other and hoped they didn't snag the corpse and bring him back on board.

They kept their eyes alert for a place to establish a marine station and when passing Puffin Island they spotted an abandoned signalling station now sadly 'destitute of doors and windows'.

The scientists came ashore there for the first time in 1887. It was a cloudless spring day with the sea blue and calm and the distant indigo hills of Wales glistening in the sun. Had they seen it in more typical conditions, with the waves assaulting the house, they would never have landed. They slept in hammocks and shakedowns on inverted tables. A team of workmen took only three weeks to clean and paint the interior, make shelves and benches and replace over a hundred small panes of glass in the big bay window.

The academics also mucked in. Herdman describes how 'A distinguished Welsh professor, who came gaily attired in tennis flannels, worked nobly as a beast of burden, until he undertook to carry a two-gallon tin of red paint, wrong side up, on his head. When he lifted it off, the lid remained on his straw hat, and the paint poured down his back like a gory cataract. We cut him, as well as we could, out of the mingled paint and flannel, and gave him an empty coal sack to wear till his luggage was landed.'

Extending in front of the laboratory like an apron was a tumble of rocks and ledges pitted with pools. Under the rocks were sprays and clumps of small animals. Nothing was what it seemed; the anemones pretended to be flowers, tube-dwelling worms sprouted anemone tops, sea squirts masqueraded as sponges, a crab was disguised as limestone. One group was called the zoophytes (animal plants) because for a long while nobody knew which they were, for 'they partook of the nature of both'. It was their reproduction that eventually gave the game away since, whatever the adult looked like, the larvae retained the appearance of their relatives.

For five hundred years sages believed the long-necked species of barnacles gave rise to geese. Sir Robert May, the first president of the Royal Society had described the bird within the shell. So had Gerard in his herbal: 'I found living things, that were very naked, in shape like a Birde; in others, the Birds covered with soft downe, the shell halfe open, and the Birde readie to fall out.' Gerard considered that a graveyard of wrecked ships just up the coast in Lancashire was the nursery of all such barnacle-born geese, because the hulks bore a 'certain spume or froth which in time breeds'. This had the stink of magic, as confirmed by Caliban in *The Tempest*: 'We shall lose our time and all be turned to barnacles.'

Gerard was an amiable enhancer of the truth. When someone described to him a log washed ashore encrusted with stalked barnacles, he drew it as an upright rooted tree with barnacle fruit hanging from every branch.

In religious medieval Ireland, geese were designated fish not fowl and therefore could be eaten on holy days. This practice kept the belief conveniently alive, even though Pope Innocent III banned the consumption of barnacle geese during Lent. Eventually, even the more fanciful and peckish clergy had to concede that the uncrusty and feathered geese were not crustaceans.

The naturalists visited Puffin Island for five summers. They brought a steam launch, but it ran into a reef and sank. The rowing boat vanished one night and returned next day in pieces, the blue punt sank on the shore. The planks of the old sailing boat, the *Bonnie Doon*, were so rotten that 'every time a leak was stopped in one place it broke out in another'. A new replacement yacht was a vast improvement until it sank at its moorings in the winter gales. Even Herdman became nervous and confessed 'how it happened that we did not drown a professor or two across that dangerous tidal race, I don't quite understand'.

They published five thick volumes listing all the species they had found. Soon every creature around the island would be named and noted. Herdman began to consider moving to a site that was both richer and easier to get to. His dredging expeditions took him to the Isle of Man. One haul contained no less than 156 species, eight of them never seen by scientists before. He determined to establish a new laboratory on the island, at Port Erin, a place that was to play a major role in my life.

The Song of the Dredge

Ballaugh, Isle of Man

When I came to the Port Erin Marine Laboratory that Herdman founded, I was the least famous marine biologist ever to work on the Isle of Man. In the library was a bust of the most famous one, Edward Forbes. He was born on the island in 1815, and until he was twelve was too delicate to go to school. The only subjects that caught his interest were drawing and natural history. The crowning achievement of his youth was to produce an 'elaborate drawing of a slug'. When his grandmother saw him grubbing beneath a hedge for snails, she said, 'I believe the whole Isle of Man cannot save this boy from being a fool.' She feared that he would end up on the streets selling slug sketches to the tourists.

Edward went off to London to study art. Within four months even he knew he would never make an artist, so he dutifully went to Edinburgh University to study medicine. It took Edinburgh four years to decide he would never make a doctor. Charles Darwin would also drop out of the Edinburgh medical school, which was in danger of its rejects being more famous than its graduates.

Meanwhile, Forbes had become addicted to dredging. He even wrote a 'Song of the Dredge':

> Hurrah for the dredge, with its iron wedge,
> And its mystical triangle.
> And its hided net with the meshes set,
> odd fishes to entangle.

He was never to be a lyricist either.

His first dredging expeditions were off Ballaugh, his family home on the Isle of Man, and his finds began to get him noticed. At the age of twenty-six, his first book, the delightful *History of British Starfishes*, was published. Here is his encounter with the fragile starfish, *Luidia*, which shivers itself to bits when caught:

> *Luidia* came up in the dredge, a most gorgeous specimen . . . Whether the cold air was too much for him, or the sight of the bucket too terrific, I know not, but in a moment he proceeded to dissolve his corporation . . . In despair, I grasped at the largest [fragment] and brought up the extremity of an arm with its terminating eye, the spinous eyelid of which opened and closed with something exceedingly like a wink of derision . . . It is now badly represented in my cabinet by an armless disc and a discless arm.

Forbes was an irrepressible enthusiast. On discovering a well-used spittoon was filled with absorbent shell sand, he emptied a sample into his handkerchief for later examination beneath the microscope.

His big chance came in 1841 when he was invited to join HM surveying ship *Beacon* on an eighteen-month hydrographical cruise in the Aegean. He found the Greeks 'almost as interesting as the shellfish that live on their shores'. This was meant as a compliment.

He dredged successfully at a depth of over 1,300 feet, far deeper than anyone had done before. But he noticed that the fauna became sparser and sparser with increasing depth and predicted that below about 1,800 feet the seas would be devoid of life. He called this dead deep the azoic zone. He was unlucky, for in the stagnant depths of this corner of the enclosed Mediterranean living creatures can indeed be sparse, but elsewhere, however deep the ocean goes, life follows.

Twenty years later, Forbes's theory should have sunk without trace when a faulty telegraph cable off Sardinia was retrieved from 6,000 feet encrusted with living organisms. But doubters claimed that the creatures had merely 'convulsively embraced' the cable as it was brought to the surface.

Forbes abandoned his hypothesis and predicted that the greatest discoveries would be made in the deep. But the reluctance of many naturalists to let his theory go had far-reaching consequences. In 1869, HMS *Porcupine* was dispatched to the edge of the Atlantic abyss off south-west Ireland to probe for life. The biologist in charge, Charles Wyville Thompson, soon recognised the inadequacies of the dredge and came up with a novel solution. He decorated the dredge with half a dozen of the mop heads used for swabbing the decks. 'The result was marvellous. The tangled hemp brought up everything rough and moveable that came its way, and swept the bottom as it might have swept the deck.'

They sampled at a depth of 14,600 feet, ten times deeper than Forbes had done, and found that 'life extends to the greatest depths and is represented by all the marine invertebrate groups'.

The excitement generated by the *Porcupine* cruise made it feasible to fund the *Challenger* expedition, which Thompson also led. It cruised the world's oceans for three and a half years and laid the foundations for modern oceanography. The *Challenger* sailed eighteen years after Forbes's death, but his mistaken theory had given birth to deep-sea biology. And that was not the only impact he had.

His early deep dredgings contained many creatures never seen before, and others known only from fossils. His fame spread and within a few years he was appointed Professor of Botany at King's College and chief palaeontologist to the Geological Survey.

He was one of the first to recognise that the biography of the fauna and flora of a region could only be understood with reference to its geological history. He also argued that fossil assemblages had probably lived under more or less the same conditions as similar communities still living. So we could infer the climate of past times from the types of fossils found; corals indicated warm waters whereas Arctic species signalled chill interludes. The climatic history of the world could be revealed.

Even as a child, Forbes had discovered that some Manx rocks were studded with fossil corals and creatures that had once trundled across the floors of tropical lagoons. The remains of tropical animals had often been found in Britain, presumably washed there by the biblical deluge. But maybe not, for in 1821 in a cave high on the North Yorkshire moors a menagerie of tropical beasts was found, including three hundred hyenas, hundreds of crunched bones and hyena droppings. It was the messy aftermath of a thousand dinners, not the high-tide jetsam of the flood. If tigers, rhinoceros and hyenas once roamed around in Yorkshire, it must surely have been *much* warmer than when I went to Scarborough for my holidays.

There was a local legend that long ago a great deer had fallen into a bog near to Forbes's mother's family home on the Isle of Man, and when someone dug, they uncovered the first complete skeleton of the extinct Irish elk, a huge beast with extravagant antlers up to thirteen feet across. It was only 9,200 years old and one of the last to survive, for they became extinct in Ireland long before they died out on the Isle of Man. Later, it would become a fragment of the fossil evidence that convinced Forbes of the former existence of land bridges linking now isolated places. He claimed that the submerged continental shelf was formerly dry land and a bridge over which animals had come to the British Isles from Continental Europe. Even today, over the submerged Dogger Bank in the middle of the North Sea fishermen sometimes net spear points used by neolithic hunters, and the bones of the wild oxen they felled.

Forbes had invented what we now call biogeography – the study of how organisms are distributed. This may seem a small achievement, but it was to spark two of the great revolutions in the way we regard the natural world. The two leading revolutionaries were both called Alfred.

Alfred Wegener was a pioneer balloonist, meteorologist and dedicated fieldworker. It would cost him dear for he froze to death on his third expedition to Greenland in 1930. He had noted that continents now widely separated shared similar geology and, for example, the coal measures in North America and Europe were identical. Moreover, they must have been formed in tropical swamps, so both regions had once been much nearer to the equator. He found that those areas whose geology matched also had many types of animals and plants in common. Forbes had stated that many organisms could only have migrated from one continent to another if some parts of the oceans had once been land. Wegener went further and asserted that all the continents had been one and over geological time had drifted apart to occupy their present positions. His theory of continental drift was at first dismissed because no one could conceive of a mechanism sufficiently powerful to move continents across the globe as curling stones glide across ice.

It was not until the Cold War that the threat from Russian submarines stimulated renewed interest in deep sea and oceanographers discovered evidence that the sea floors were spreading outwards and carrying the continents apart on giant conveyor belts. India has crashed into Asia, crumpling the land to form the Himalayas; Africa is creeping up on Europe and slowly closing the gates of the Mediterranean at Gibraltar. As the Atlantic widens, the Americas are sliding away from Europe; it was easy for Columbus to discover America – it was eighty feet closer in his day.

The other Alfred was Alfred Wallace, a naturalist who made his living by selling animal specimens that he had caught in the tropics. While exploring a tiny island in the East Indies in 1857, he pondered why its fauna should be so similar to that of New Guinea and Australia to the east, yet so different from that of Java and Borneo to the west. If, as everyone assumed, God had designed species specifically for particular landscapes, surely New Guinea should have the same fauna as all those westerly islands with similar dark and dripping jungles. Conversely, if kangaroos were suited to the arid plains of Australia, why were they also found climbing trees in the damp mountains of New Guinea? Surely they could not be 'designed' for both. Following Forbes's teaching, he concluded that New Guinea and Australia had once been connected. But he claimed much more than that: 'new species have been gradually introduced into each, but in each closely allied to the pre-existing species . . . The species which at the time of separation were found only in one country, would, by the gradual introduction of species allied to them, give rise to groups peculiar to that country.'

By the word 'introduction' Wallace did not mean brought in from outside or created by God. He stressed that these new species were closely related to pre-existing ones and it was widely accepted that 'sports' and varieties arose naturally. 'Why,' he argued, 'should a special act of creation be required to call into existence an organism differing only in degree from another which had been produced by natural laws?'

Such thoughts were heresy. He was saying that they had evolved from existing species without the need for divine intervention.

While Wallace scaled mountains in the East Indies, half a world away, in his study in Kent, Charles Darwin was laboriously trudging through mountains of data that would, he hoped, convince the scientific world that the evolution of living things had occurred and could be explained by his theory of natural selection. He would have to be cautious for he knew that the mere juxtaposition of the words 'Origin of species' would be revolutionary. Amassing the evidence would be a long job and he was anxious not to be rushed into premature publication.

Darwin had attended a talk by Wallace and had said to him that they were 'thinking along the same lines'. This didn't lessen the shock when two years later a letter from Wallace exploded through his letter box. It was the draft text of a scientific paper, entitled *On the Tendency of Varieties to depart Indefinitely from the Original Type*, in which Wallace outlined *his* evidence for the evolution of species. Darwin, who had already written 250,000 words of *The Origin of Species*, had been pre-empted; he had lost priority on the most important idea in the history of biology. He was desolate and must have been tempted to burn the letter, to pretend it had never arrived; indeed, some scholars claim that he held on to it for a month, pondering what to do, before forwarding it to Sir Charles Lyell for publication, as Wallace had asked him to do. Lyell knew of Darwin's theory and had earlier tried to persuade him to publish at least an outline lest he should be pipped by the post. Now he suggested that if Darwin promptly produced the outline, his paper and Wallace's could be presented simultaneously at a meeting of the Linnean Society. Our view of the natural world would never be the same again.

Forbes did not live to see the great theories that sprang from his ideas on biogeography. In 1854, four years before the historic joint presentation at the Linnean Society, he was appointed to the chair of

Natural History at the University of Edinburgh. The senate must have failed to spot that he was their failed medic in disguise. He was on the brink of greatness, but taught only one course at Edinburgh. That summer he fell ill. An infection contracted abroad had damaged his kidneys and he died. Forbes was thirty-nine.

He was one of the most original thinkers of his generation, an immensely serious man with a well-developed sense of humour. He named a pretty little jellyfish *Lizzia blondina*, perhaps in honour of a blonde called Lizzie he knew in Ballaugh on the Isle of Man.

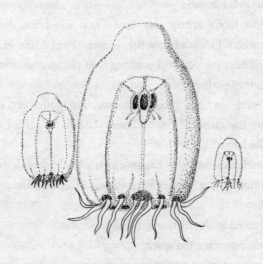

All at Sea

The Atlantic Ocean

When I got my degree in 1963, I was put forward for a prestigious Development Commission studentship to do research. The interview panel with the grant in their gift was a formidable array of knights, the leading marine biologists of the day. They asked me dozens of questions to which I didn't know the answers. Eventually, in desperation, they fell back on 'Are you working over the long vacation? At a marine laboratory perhaps?'

'No,' I said, on safe ground at last. 'I'm going on a diving expedition to Lanzarote. In the Canary Islands,' I added for the geographically inept.

'Lanzarote!' exclaimed Sir Frederick Russell. 'A wonderful place for siphonophores. You have, of course, read Haeckel's *Monographie der Siphonophoren*?'

'Only glanced at the pictures,' I said modestly.

They didn't give me the grant.

There were ten of us on the expedition: three geographers and a geologist, four medical students masquerading as anthropologists, a botanist and me – a one-man diving team who would collect and identify the underwater flora.

We sailed from Southampton on a Spanish tub, the SS *Begona*.

The ship's officers looked smart in their tropical whites but the gold braid that festooned their caps turned out to be gilt. I lost faith when I spotted the captain dissolving anti-seasickness Quells in a glass of gin before we faced the deep swell of the Atlantic. My doubts were confirmed when we reached Vigo and steamed in circles for an hour. 'We fit a new compass,' a steward confessed. 'An' we try him out.'

We then crossed the Bay of Biscay, which has a sinister reputation as a place of storms. The truth is that storms are more common on the west coasts of the British Isles than here, but it is the first encounter with the Atlantic for passengers from Britain who have not yet developed their sea legs. Their queasy unease has forged the legend of the violent Bay of Biscay.

The Atlantic Ocean is a magnet for myths. It is the home of the Bermuda Triangle, that infamous swallower of ships to those who will swallow anything. To the south-west of the 'Bay' is the Sargasso Sea, a real place where a huge gyre of currents encircles an area the size of Greenland, trapping floating seaweed. Here, it has been said, web-footed tribes of mermen live beneath the wrack. Columbus thought he had sighted land only to become becalmed for a fortnight in a wilderness of weed, while below deck his crew discussed mutiny. It was claimed that the malicious plants 'surround the ship in such quantities as to retard its progress'. Sailors spoke in hushed tones of hulks gripped by weed, their crews long dead from thirst and madness. But the floating rafts of *Sargassum* weed are only thirty feet across at most and couldn't possibly impede the progress of a ship coaxed forward by a breeze. And therein lies the problem; the Sargasso Sea is almost forgotten by the winds. Sailing ships are always likely to linger there a while.

As often happens, the folklore obscures the true wonder of the place. It is estimated that over 10 million tons of *Sargassum* drift within the gyre. Curiously, the usually fecund plants have forgotten how to become fertile. Instead, they rely on fragments breaking off and growing into independent plants identical to their parent. They are cloning themselves and therefore may live for ever. So if you are ever becalmed in the doldrums of the Sargasso Sea, lean over the side and hook up a plant – it may have been last handled by Columbus.

The weed nurtures a sparse but unique fauna, including wrack-tinted crustaceans and fish, their bodies extravagantly adorned with leaf-like lobes so that they are indistinguishable from the weed. The *Sargassum* anglerfish angles with a lure dangling invitingly from its nose and has prehensile fins so that it can cling to the weed. There are also visitors such as loggerhead turtles that hatch on the Florida coast and spend their adolescent years in the Sargasso Sea.

The freshwater eels that abound in the rivers of Europe and the eastern seaboard of America never breed there. When they are eight to ten years old they change into silver livery, promptly swim down to the sea and are never seen again. In the spring, trillions of tiny juveniles appear in every estuary and swim upstream where they grow into adult eels. Where do they come from? Astonishingly, the eels breed only on the bottom of the Sargasso Sea and then die. It takes their tiny finless larvae three years to drift the 2,500 miles back to the European rivers from which their parents departed. Nobody knows why somewhere nearer to home won't do for breeding. Clearly, the Sargasso Sea must be a very special place.

We have this story because in the 1920s a Danish oceanographer who had wisely married the heiress to the Carlsberg brewery was able to finance expeditions to trace the path of the migrating eels. However, the mystery is not quite solved, for no one has yet retrieved an adult eel or a fertilised egg from the bottom of the Sargasso Sea. Eels seem to hold on to their secrets. A young research student dissected over four hundred eels, but failed to locate their reproductive organs. He gave up and went off to study human psychology where evidence of sex would be much easier to find. His name was Sigmund Freud.

The placid Bay of Biscay had lulled us into a false sense of stability, but there is always a wave lying in wait to wet the unwary. Our cabin was the black hold of Calcutta. It was far too small for six bunks and the tiny fan did nothing to combat the heat. I had the bunk beside the open porthole and was finally cooled by a wave that came in and drenched me. We were sinking! I leapt to the window. There was nothing to be seen except a silk-smooth and sleepy ocean. Nobody else had even woken up.

I could feel the tremor of the engines beneath my feet, and hear it clank and jangle like a Trinidad steel band. It *was* a steel band. I got dressed and burrowed to the bottom deck of the boat where a dance was in full swing. Every face was black and smiling at the prospect of heading home to the Caribbean, the ultimate destination of the ship. The men were all tall and muscular and the bright-eyed women wore pink or yellow cake frills for frocks. I was welcomed, given half a pint of rum, and an ovation when I played the drums. It was a far cry from the sedate dances for the white passengers on the decks above, where foxtrots were enjoyed in slow motion because the record player was running too slowly.

At Las Palmas we changed ships. The inter-island boat was the elderly *Gomera*. She was already fully loaded, so most of our luggage would have to follow on tomorrow's boat. The deck was crowded with people and poultry; somewhere in the throng a pig grunted from both ends. Bullocks were being swung aboard in a sling. There were fewer livestock on the Ark. As we had no cabins we would have to share the deck with this menagerie.

It was the filthiest ship afloat. If they had scraped and scrubbed the deck she would have floated a foot higher in the water. The lifeboats were rusted to their davits and cracks in their hulls had been filled with paint. I was not reassured by the sight of a rat leaving a *floating* ship.

The stern was full of soldiers on their way to join a garrison on Fuerteventura. They showed me photos, not of wives and girlfriends, but of tarts in Madrid who did wonderful things for three hundred pesetas.

The only lavatory on board had clearly been blocked for some time, although this didn't seem to adversely affect custom. I decided I didn't need a pee after all, and anything more meaningful would have to wait until I returned to England.

As the sea got up, the *Gomera* struggled to keep on course. In the moonlight the deck was a silvered cemetery of bodies, the women shrouded in thin cotton shifts. Soaked and shivering, some whispered eerily among themselves and kept their babies snug beneath them. I sat beside an open hatch where warm oil-laden air rose from the engine room. In spite of the noise, and the smell of clogged latrine, I eventually fell asleep and awoke with the sun.

I went to the prow to look for land. Suddenly, a flying fish broke the surface and burst away from the bow clipping the tops of the swell like a skipping stone. Then there was another, and another, an entire squadron of these supreme ocean gliders. The best of them were making flights of up to 170 feet. They launched themselves into the air, with their fins outstretched like wings. Initially, they left their trailing tails beating in the water at sixty strokes a second to provide continued thrust until they reached sufficient speed and lift to glide away. They enjoyed the best of both worlds: using dense water to provide the propulsion and then up into the thin air with its lower drag.

The joy of being a marine biologist was that the flying fish were not just a visual treat, but also a demonstration of hydro- and aerodynamics. And understanding how the trick was done only added to the magic.

I gazed into the indigo ocean and watched the bow sever the calm water. It seemed to leave behind a plume of pale smoke on either side, but where it sliced open the sea, the colour was dazzling. Now I understood the meaning of aquamarine.

It was some while before I looked up, and there, arid and ominous, Lanzarote rose from the morning mist as if it were being discovered for the very first time.

Stars and Starfish in the Sky

Los Ajaches, Lanzarote

Arrecife, the capital of Lanzarote, was in sight for two hours before we docked. The journey had lasted twenty-two hours.

Arrecife was, in those days, a one-hotel town and it was more than enough to accommodate the few visitors that came. Trippers might fly over for the day from Las Palmas, have a camel ride around the Montañas del Fuego, but few stayed the night. They failed to discover that the hospitality here was even warmer that the craters.

We were met on the quay by Don Mariano López, a local dignitary and ex-mayor, who kindly took visiting expeditioneers under his wing. A donkey cart took away our luggage and we were whisked away to an eight-course lunch topped off with slices from the world's biggest melon, coffee and cognac.

We had been allocated an empty house on the edge of town. Our arrival caused such interest among the locals that a policeman had to keep the crowds back. The girls didn't come out, but displayed themselves ostentatiously at the windows, seeing and being seen. The house had no water supply so the neighbours brought bucketfuls and Don Mariano provided a crate of mineral water.

The night was oppressive. The heat, and the rattle of cockroaches in our plastic bags, made it unbearable indoors so I slept in the yard.

In the velvet darkness the huge, hemispherical sky was cluttered with stars and the luminous smoke of the Milky Way. I had never really seen the night sky before, only the sodium-jaundiced glow of cities.

Next morning we went out for breakfast. I asked, in what I thought was Spanish, for something typical of the region. I was served bacon and eggs. At least I think so. They knew the recipe – fill a swimming pool with hot oil, then submerge a dozen eggs – but what adorned my plate was a slab of translucent fat without a trace of meat and two green-eyed objects staring out from a mask of congealed white rubber.

Our impatient geographers left for the south of the island, taking half our supplies. We would be all right so long as the crates left behind in Las Palmas arrived in the next couple of days. They didn't, of course.

While waiting for the missing stores to turn up, we explored the local shores and then took a siesta. This was a wonderful escape from book learning. There is a Spanish saying: 'There is nothing nicer than doing nothing all day, then having a rest afterwards.' When abroad it is important to fit in with the locals.

By dusk an onshore breeze fanned the town, the flies settled down for the night and a swarm of shoeshine boys took their place. Arrecife came to life. The darkness was parted by the glare from the barbers' shops and the dim glow from the grocers'. Elbows and faces protruded from every window. Families strolled around the park and clutches of girls went round in fours for safety. The boys also went in fours, to protect themselves from the girls.

The girls were brown and beautiful and very different from those back home. 'Do you like to see my feet?' one asked. Well, it was a start. She slipped off her shoes to display her dainty toes – all twelve of them.

We ended up in the Bar Janubio where everyone chatted and

laughed and the waiters whistled. A dead-and-alive band played Iberian versions of pop songs, Spanish ballads and novelty numbers. We all sang and smiled a lot and felt that all the evenings of our lives should be like this. We were, of course, very drunk.

Don Mariano held a party for us. His wife laid on a feast of goat's milk cheese, maize-flour bread, shoals of sardines and mullet followed by sweet sugar cake and all accompanied by a dry white wine from their own vineyard. His daughters served us. They were in their mid-teens but, as Senora López warned, here they marry when only sixteen or seventeen. 'It's the climate,' she said, despairingly.

Afterwards, we were taken to a dance at the community centre. Don Mariano alternately forbade his girls to dance or thrust them on to the floor when they coyly refused an invitation. They were crazy about the twist so I asked him if one was allowed to dance with me. 'Of course,' he said, as if foreigners were no threat. Even so, as she got up she crossed herself.

Every day we went to the dock hoping to pick up the missing supplies and on the eighth day they arrived.

I set off for Femés in the south with relief provisions for the geographers. They had intended to camp, but Don Mariano had arranged for them to stay in the deserted church.

South from here was only arid *malpaís*, badlands. There was no ash or dust, the entire region had been submerged in a solidified sea of black lava. The land had turned to stone and looked as bare and forbidding as it did 230 years before, when the lava tide first flowed. Under the incandescent eye of the sun, I crossed this awful desert to climb up to the high beaches.

Sea level has not been constant throughout time. Seventy-five million years ago the ocean's surface was 1,300 feet higher than it is today. Since then it has progressively fallen, but there have been fluctuations along the way. During the glaciations, so much of the world's water was converted into ice over the land that sea level fell over 320 feet. Thousands of years ago early man painted the walls of caves that are now visited only by divers deep beneath the ocean.

We are now in the lukewarm wake of an ice age. The alpine glaciers and the polar ice caps are still in retreat. But in the truly warm periods that lay between the frigid ones these ice stores melted away to almost nothing and ran into the sea. If all the ice should melt again, the sea would rise by 260 feet and almost every city in Britain would be drowned. The land would be reduced to an archipelago of little islands where only the former peaks protrude above the waves.

Lanzarote was never closer than 1,800 miles to the ice, but it felt the wash of the rising and the falling of the ocean. In the south, the desolate Los Ajaches uplands bear the scars. Here there is a series of six wave-cut benches where the sea repeatedly rested while falling from 180 feet above the present shore. Each bench is littered with pebbles rounded by the sea, scatterings of ancient seashells and even the occasional starfish. These are the ghost beaches of long-lost tides and a reminder of the uncertain balance between the land and sea.

Slitherers
and Skeletons

El Golfo, Lanzarote

The expedition's botanist was Martin Hurst, a lecturer at Liverpool University. He was sharp-featured and sharp-witted and, he assured me, prematurely grey. 'I've had a brilliant idea,' he said when I returned to Arrecife. I became nervous when Martin had brilliant ideas. 'You know those big tricycles the delivery boys have in town?'

'Yes . . .'

'Well, I've hired one to take us across the island. We can pedal to wherever we want. We'll be free as –'

'Stop right there. How many tricycles with a bloody great platform up front finish the Tour de France?'

'That's the beauty of it. The platform will carry all our gear.'

The contraption was five times heavier than our luggage. It had three gears, one of which worked. With the platform fully loaded, we took turns to pedal up the inclined road that led out of town. The incline soon became a slope, the slope a gradient, and the gradient grew and grew until it was indisputably a hill. Beyond the first hill was another and then another and then . . . we turned round and sped back to town. The brake kept locking the back wheel leaving snaking rubber marks on the road.

We took the postal bus instead. It was a rectangular tin box constructed so that no piece was securely attached to any other. The independent movement of so many parts gave the feeling that at any moment the vehicle would disassemble into its constituents and leave us sitting on a flatbed truck in the open air.

The driver, with a tiny pillbox cap perched uncertainly on his head, sat in a well at the front with the huge gear lever emerging a metre behind him and stretching horizontally forward as if searching for a hand. He had to change gear behind his back.

It was the noisiest vehicle ever built, rattling along the crushed ash road, past the three hundred still-hot craters of the Montañas del Fuego. The lava flows had emerged over six years of continuous eruption in the eighteenth century and covered a quarter of the entire island. In the middle of this desolation, the driver stopped the bus, got out and vanished. I found him fifteen minutes later beneath the back axle having a smoke.

Yaiza was a small, whitewashed place with a pretty church bright against the black lava fields beyond. The plaza was also the lid of the town's reservoir. We set off to walk the four and a half miles to the coast. When the sun got up, the road was almost too hot to tread. At last we made it to the coast and the cool breeze from the ocean.

We camped at El Golfo where an arc of cliffs, in seven shades of chocolate, embraced the beach. They were the remaining half of a volcanic crater; the sea had consumed the rest. The cliffs sheltered the bay from the east, but today, as always, the fierce wind was from the west and had crossed the entire Atlantic Ocean just to uproot our tent pegs and rattle our flaps.

Low down, the black beach was littered with the empty carapaces of carmine crabs and there were long pools in which stranded fish waited for the tide to return. They were designed for dinner. I wedged a hand net behind a low dam I built at a constriction between two pools. With the maximum amount of splashing, I herded the fish towards the dam and, like apprentice salmon, they leapt over it and into the waiting net.

I returned triumphantly to camp very much the hunter-gatherer, but Martin had already cooked sausage and mash (one part sausage to thirty parts instant potato). Reluctantly, I returned my delicious fish to the pool and attempted to excavate the mash.

'Yummy,' said Martin.

Fortunately, the meal was rescued by a local family who had heard of our arrival and came down to the beach with a plate of fresh figs and a bottle of wine. We sat together on the sand bar that separated the ocean from a viridian lagoon, even saltier than the sea.

'*Agua estancada*,' said the mother of the family.

'*Sí*,' Martin agreed. 'Outstanding.'

But *estancada* means 'stagnant'.

They were typical of many local families who fished for three months during the summer, but in winter moved inland and cultivated vines and sweet potatoes out of the wind in deep funnels lined with volcanic ash to absorb the dew. That anything could grow in this arid wilderness was a miracle. They had also cultivated cochineal beetles on *Opuntia* cactus. It took 70,000 bugs to make a pound of cochineal yet in 1860 the annual harvest had been worth $US 1 million. Then the development of aniline dyes ruined the market. Impenetrable thickets of the plants remained, some still blighted by the white fuzz of hairy beetles. In the 1920s, *Opuntia* spines had been the top-of-the-range gramophone needles for getting the best sound out of your 78s with the least amount of record wear. Then someone invented microgroove LPs and thereafter the cacti could keep their spines for scratching tourists.

The family's eldest son, José, had studied science on the mainland and cartography in the army, but returned to fish and farm with the family.

'Would you like to come fishing tonight?' he asked.

*

It was almost midnight before we saw an eerie glow advancing over the rocks. José was carrying a blazing brand, like a seventeenth century smuggler. It was a long metal cone filled with meths, which fed a rag wick at the top end. The flickering light reached to the very top of the cliffs.

We trudged off together to search the pools on the lower shore.

'What are we looking for?' I asked.

'*Pulpo*,' he said, lowering his brand until it almost touched the water.

A small octopus emerged from a cavity beneath a rock like ectoplasm issuing from the mouth of a medium and slid over the boulders with an easy ooze that was impressive. It became transfixed by the light so I picked it up, and immediately regretted it. The octopus has a flair for inciting repugnance. It changes its texture and colour to impersonate boulders or weed or a slither of slime without substance. Even slime doesn't take this as a compliment. Imagine a half-inflated football that can easily squeeze through a keyhole then reform into a tangle of tentacles and suckers. If you are afraid of snakes, it has eight of them. Quick as a whip, its arm embraced me, then another and yet more. They could do eight unpleasant things at once. Pulling them off felt as if they were taking my flesh with them. I was lucky; when it mates, the male pinions the female in an hour of foreplay or, in this case, eightplay. It insinuates a sly tentacle underneath the female's 'skirt' to slip her a capsule of sperm. Sometimes he squeezes her gills so that she swoons from lack of oxygen. This rape shouldn't be necessary, for it is possible to stroke an octopus in the appropriate place and watch its eyes glaze over.

The octopus is an Einstein among invertebrates. Even hardened researchers recount that they are quick to learn, they come to be fed and their gaze follows you around the room. At least, one of their eyes does because some are right-eyed, others left-eyed. The structure of the octopus eye is almost exactly the same as ours and it is difficult to gaze into it without thinking that a human being is trapped inside. It has the look of sad acceptance of someone transformed into a bag of jelly by witchery and condemned to unhappy octopusdom for ever.

There was a greater unhappiness to come. José threaded it on to a stiff wire attached to his belt. He inserted the wire into its mouth and out through the top of its head. It stared at me in sad surprise, as if asking what was I going to do about it. Eventually, seven or eight were skewered on the wire, stacked one above the other, still half alive and pathetically waving goodbye. It reminded me of Pepys' description of a man who was being hung, drawn and quartered: 'He looked as cheerful as anyone could in his situation.'

A famous spearfisherman once observed: 'When the octopus is dead, which he rarely is, he doesn't look much better than he does when he's alive.' But was it my imagination, or was the look when dying quite different? All their suffering was for nothing, for the rubbery octopus can be about as tasty as a tyre. No matter how hard you pound the meat, or how long you boil it, it remains a huge novelty eraser.

On the way back, José said he would show us something special and led us deep into a cave. It was a typical volcanic tube. Long ago, the hot lava had run all the way through leaving an empty tunnel. We trudged in for over half a mile, following the burning brand. Shadows on the walls swung dramatically and lurched out towards us as we passed. We hardly spoke, for every word skittered away from us to get lost in the blackness.

'*Aquí*,' said José at last, and held the brand down to reveal an entire human skeleton half buried in the brown dust that carpeted the floor. The skull grinned at us with abnormally large teeth.

'He is here a long time,' José explained. 'I find him more than twenty years ago.'

'There was no trace of clothing?'

'No. *Guanche*, I think.'

In 1336, when Lancelotto Malocello first came to settle on Lanzarote and give it his name, he found indigenous cave-dwelling natives, the *guanches*. They were naked except for hide cloaks and tattoos. Women were in such short supply that each was allowed three husbands who would operate a strict rota, with each spouse having a month on, so to speak, and two off.

The men were prodigious fighters and honed their skills in games, with contests including wrestling, which retains its popularity to the present day. Such was the pride of the ancient warriors that a loser often threw himself to his death over a cliff to prove his valour. Not to be outdone, the winner sometimes followed him.

The origins of the *guanches* are a mystery, but they mummified their dead in the fashion of the ancient Egyptians and there are even the remains of small step pyramids in the Canaries. It is probable that the *guanches* came from a Berber stock that had earlier given its beliefs and customs to the Egyptians. Indeed, Cleopatra's daughter married a Berber chief at about the time the Canaries are thought to have been colonised. Most likely, the *guanches* derived from undesirables exiled without boats and therefore with no possibility of returning to the mainland. A local legend claims that 'God placed us on this island and then forgot us'.

The museum in Gran Canaria is full of *guanche* skulls variously perforated and cleaved to let in Christianity from Spain. Many of those that survived conversion died of disease or in battle, for their slingshots and wooden javelins were no match for European weaponry.

Perhaps the corpse we had found was indeed a *guanche*. On other islands hundreds of mummies had been discovered laid out in caves, but there was no sign that our body had ever been mummified. So maybe he was just an unsuspecting visitor who had been lured down into the cave and murdered.

The night was without a moon and the stars forgot to shine. A solitary candle made our tent a tiny ochre pyramid in the immense darkness. The wind had died leaving an eerie stillness broken only by the harsh breathing of the waves against the sand. Then something dived low over our tent. Time and time again it swooped down with a rattle of wings and an unearthly elastic scream.

'What the hell is that?' Martin murmured so as not to attract its attention.

'A shearwater?' I said reassuringly.

That night we both dreamed of dark and terrible tunnels, and the skeletons of lost men.

Sea Meadows

Caleta del Sebo, Graciosa

Arrecife was the major fishing port of the Canaries. A couple of dhows with their off-the-shoulder sails were coming in over the horizon from Africa, but diesels propelled most of the boats in harbour. They were wide-bellied and swoop-bowed, built to ride the big Atlantic swell.

We decided to catch a lift on Don Mariano's thirty-four-foot fishing boat to Graciosa, a small island to the north of Lanzarote. Martin had felt queasy on all the voyages so far, so I tried to reassure him. 'On larger vessels,' I said authoritatively, 'the amplitude of the oceanic swell sets up a slow roll that induces a predisposition to nausea. The happy bounce of a small boat is quite different.'

He looked dubious.

'Trust me.'

Martin spent the entire voyage the colour of custard and muttering, 'You bloody liar.' The worst things about seasickness are the feeling you will die, then the realisation that you won't.

Our team of budding anthropologists had been on Graciosa for a couple of weeks and were now local heroes. Although far from qualified, they were the only thing approaching doctors on the island and had worked wonders. In their first week they had deflated an angry sty, restored a dislocated shoulder and cured a septic foot.

They were modest about their training and assured me that almost all diseases are either fatal or the patient recovers. In either case medical interference just prolongs the agony. 'We're not taught to cure diseases,' they said, 'but to discover new ones to ensure the future of the profession.' Nonetheless, I was envious. Even tyro medics were *useful*. Most biologists long for the day when some modicum of their knowledge will become relevant to everyday life, or a chance to air that witty aphorism about mating sea slugs. There is a botanist's dream that the barbecue cook apologises for the slow sausages. 'It's this rotten charcoal,' he complains. The intrepid botanist whips out his magnifying glass, scans a briquette and astonishes the crowd with 'This isn't oak. It's an inferior strain of ash! Although,' he confides to an adjacent blonde, 'to the layman's eye, the xylem is somewhat similar.'

Marine biologists await in vain the magical moment in a restaurant when a diner asserts that the scampi is really slices of dogfish in disguise, and the waiter cries: 'Is there an ichthyologist in the house?'

The medics' anthropological work wasn't going too well. They were taking lots of physical measurements, such as cranium diameter and nose length, because it was easy to persuade the locals to submit to that. They had also gathered over two hundred tiny samples of blood for grouping, but 90 per cent were from women. 'Look, it doesn't hurt,' they said, stabbing their own fingers with a small medical lance. The men nodded, but didn't proffer a thumb.

The medics thought they were on to some old *guanche* anthropological traits, for many of the locals had buckteeth. Later, they discovered that they also shared the same surname. Even the original European settlers, who were not agriculturally inclined, scattered their seed widely. The Spanish colonised most of the islands (and to this day peninsular Spaniards are called *godas* – Goths), but the French came first and concentrated their efforts on Lanzarote and nearby Fuerteventura. They were led by a Norman knight, Jean de Béthencourt, and now every taxi driver and grocer on Lanzarote seems to be called Béthencourt or some variant of it. Those old fellahs really knew how to conquer.

It was at Graciosa that they first landed in the Canaries in 1402. Here they killed their first *guanche*. On Lanzarote they cleared the ground of natives overlooked by earlier slave traders. The commander boasted that all the fighting men were slaughtered and the women and children baptised and enslaved.

Graciosa had changed little in over five hundred years. It was still a place of sand and silence, a small fishing community rarely visited by outsiders. It would, however, soon have the ultimate status symbol of Canarian towns; 'We are building a cemetery,' a local boasted.

Some old fishermen hauled their boat laboriously up the beach as if it were a great fish they had wrested from the ocean. Younger men came to give a hand. They all wore narrow straw cloches and denim clothes bleached to ten shades of faded forget-me-not. For contrast, one young boy's shirt was a kaleidoscope of checks and tartans, perhaps twenty patches from as big as half a sleeve to as small as a postcard. There was no way to tell whether any of the original material survived. These clothes might be truly immortal – everlasting jackets, and trousers that would walk on long after their first owner had passed away. The women wore gingham frocks with a flounce of petticoats beneath, and an apron above. They were mushroomed by wide-brimmed straw hats and stayed paler than most women in northern Europe.

Graciosa was a golden desert. The sky may be populated with speeding clouds, but they watered fields elsewhere. Little grew here and much had to imported to feed the eight hundred inhabitants.

Camels were the beasts of burden and fishing was every man's trade.

The square was covered in sardines drying in the sun. A circle of children gutted them by hand, leaving the heads on as they added to the weight when they went to market. That night some of the sardines were roasted on glowing embers and we drank wine, lay back and lost count of the stars.

By day I explored the shores and collected specimens to be identified back at the house. Working at my microscope, I gathered a crowd of local urchins eager to know what I was up to. I tried to explain what was going on beneath the lens in terms that wouldn't strain their limited understanding or my even more limited Spanish.

'*Lo pongo las algas dentro el tubo.*' ('I put the seaweeds into the tube.') So far so good, but what about 'The spores are being extruded from their capsules'? '*Los niños estan salidas su casa.*' ('The children are leaving their house.') Algal reproduction was reduced to 'Mother and Father plants meet and then . . . and then . . . *olé!*'

The shores were inhabited by animals and plants that I knew only from books. A fountain of sea water gushing up from the rock was the blowhole from a lava tunnel below and the incoming tide was thrusting up through it. Around the rim of the hole a garden of stunning seaweeds flourished in the stream. But a beach doesn't need subterranean tunnels to confer mystery and luxuriance. I was to learn that every shore I would ever visit would hide secrets below the surface.

One morning Martin and I trekked across the dunes that stretched behind our house, scattering grasshoppers and iridescent turquoise dragonflies with every step. The glare from the sand was blinding and the rising heat roasted our ankles. Climbing the scree flanks of a baked hill lacerated our shoes, but from the top we could survey the whole of sea-encircled Graciosa. It was bereft of any vegetation taller than dune grass.

On the way back a stiff breeze blew in from the sea and flying sand began to sting our faces. Sand is just a heap of ground-up glass and mobile sand is frightening. It can scour all life from the land. Deserts are deserts because the vegetation has failed to hold the sand down.

On the adjacent shore there was another sandy plain underwater. When the waves awoke they moved the abrasive sand as effectively as the wind on land. It would have been a very wet desert were it not for some special plants that keep the sand in place. The seagrass *Zostera*, for example, had a more robust underground stem, like bracken. At frequent intervals, tufts of green strap-like leaves emerge from the sand to form a swaying meadow. Underwater vegetation attracted small seaweeds and hydroids to perch on its leaves and armies of tiny snails to eat them. On the sand between the tufts there was a city of snails going about their business, including a relative of the sea hare I had seen in a lagoon on Anglesey. It stood on its head its delicate shell buried in the sand and its 'foot' protruding. I assumed that like some of its kin it lay in wait to grasp passing prey.

Once upon a time these emerald meadows were abundant all around the world, and dried *Zostera* leaves were widely used to insulate walls; all the original studios at the BBC were soundproofed with them. But in the early 1930s the meadows were wiped out across Europe and North America by a mysterious 'wasting disease' of *Zostera*. Along a 5,000-mile stretch of the Atlantic coast of the United States masses of rotting plants washed ashore together with the animals inhabiting them. There was also a catastrophic decline in the numbers of Brent geese that relied almost entirely on *Zostera* for their winterfeed. Limpets are the epitome of tenacity, but a species that lived exclusively on *Zostera* became extinct.

A slime mould caused the wasting disease, perhaps because

Zostera's resistance was weakened by pollution. In most places it never fully recovered and periodically the wasting disease recurred.

Here on Graciosa, in the wash of the clean Atlantic tide, *Zostera* still thrived, along with the rich community that sheltered between its leaves. As we colonise more and more of the world's coastline and unroll a concrete carpet of roads and hotels, there are fewer of these pristine regions where nature can safely go about its business and provide oases from which one day we could repopulate despoiled regions abandoned by developers.

There are other sand binders such as *Caulerpas*, seaweeds that resemble green leaves, fern fronds or clusters of tiny grapes. They arise from creeping runners, much as strawberry plants do, and the underground cobweb of runners keeps the sand in place. Unlike all other plants and animals, which are built of millions of tiny cells each independently controlled by its own nucleus, *Caulerpa* shuns compartmentalisation in favour of tubular construction. No matter how many metres its runners spread or how numerous its upright fronds, the entire plant is a continuous hollow tube without interruption. As it doesn't waste energy manufacturing cross walls, the tube can extend at a phenomenal rate. Floating in the film of protoplasm that lines the inside of the tube are billions of nuclei, although how these coordinate its activities is hard to imagine. *Caulerpa*'s non-cellular success shows that there is another way to construct living things that works. Surprising then that few other organisms have followed suit.

My favourite plant from these submerged sands was *Acetabularia*, a formidable name for such a delicate plant. Imagine a tiny parasol perched on a thin stalk, the whole thing apparently made from pale green chalk and resembling a beautifully stylised daisy carved by a Japanese craftsman. Yet it is merely a relative of *Caulerpa* reinforced with calcium – a common ploy to make it less appetising to grazers. It also has an internal surprise. Whereas *Caulerpa* has a superabundance of nuclei, *Acetabularia* has only a single mega-nucleus at its base, which manages to supervise the growth, development and daily activities of the 'root', stem and umbrella of one of the prettiest things in the sea.

We had found our desert island, but now we had to leave. The boat back to Arrecife departed at three in the morning. Martin and I consumed what had become our standard breakfast, several coffees and a mush of Weetabix and Bemax suspended in a slurry concocted from dried milk powder and sugar. We got to the quay by 2.30 a.m., but the boat didn't. A group of women also waited, but slept beneath their mushroom hats inside an open boat.

I used the public lavatory, a sentry box with a hole in the floor, hanging over the end of the quay. The stream of urine caused a stir in the darkness below. Creatures began to luminesce and who could blame them. I stared down at them and surely some of them stared back at me.

I questioned an old fellow about the expected boat. He confirmed that the boat was coming at three. 'But it's now almost seven,' I said.

He agreed it might be late.

Fishermen's Tales

Punta Mujeres, Lanzarote

The postal bus wheezed and rattled its way up the twisting mountain road to the north of the island. It was dusk when at last we left behind the volcanic wasteland for the pleasantly palmed oasis of Haria.

We found a cheap *pensión* and the old lady in black who took our money was so bent that her nose reached perilously close to the floor. The dining room was in deep gloom, perhaps because the lukewarm stew was littered with things that were too close to diced body parts for peace of mind. The little old lady scuttled in and out of the pool of light spilt by the candle on our table, as if she were a spider checking whether anything juicy had strayed into her web. Martin didn't help by calling her '*Señor*'.

That night we fell asleep to the wonderful sound of singing from evening mass at the church, voices as soft as wisps of smoke intermingling. I felt that I had been comfortably interred and the hymns were for me.

We had come in search of soil. Martin had some clever procedure back in Liverpool by which he could analyse residues in soil and find faint chemical echoes of the plants that had once grown there. By coring deeper into the soil he could reach back in time and unravel the vegetational history of the area. The only problem was that he couldn't find any soil. There were some ash-covered fields, but what he wanted was wild, uncultivated areas. We searched the hinterland, but there was no soil to be found – it had all been swept under a carpet of lava.

A desperate Martin had come to a decision: 'I'm going off to Tenerife. They're certain to have some soil there . . . aren't they?'

'Of course. They wouldn't be seen without it.'

'What about you?'

'I'll go and stay with friends.'

I had met Emelio on the *Begona*. He had spent a year in England and loved every minute of it: the freedom, the girls, the dancing and our colourful language. His family lived in Las Palmas where his father was a fruit exporter, but they spent the summer at their holiday home on Lanzarote. Emelio had invited me to join them.

The bus dropped me in the middle of a lava field and the driver pointed me towards the coast. After a long, hot tramp, I came to Punta Mujeres, a line of small white houses with the rugged lava field behind and the rugged shore in front. It had a windmill and a stretch of salt pans like horizontal windows of water. Most of the world gets its salt from the unlimited reserves in the sea. If all the ocean's salt could be laid across the land, it would form a layer almost five hundred feet thick.

Trawling his supply of traditional English greetings Emelio said, 'You look shagged out.'

His mother was homely, his father dignified, not at all like Emelio who had the cheeky smile of a laddo who was up to no good. We met up with his mates in the local bar and drank the cinnamon-coloured wine. The tiny café was dark inside. It had white walls, but the ceiling was black, at least I thought so until Emelio tossed up a dried sardine. Where it hit the ceiling a pale halo opened then closed again. The ceiling was white, but coated with flies. We drank and sang until, in Emelio's words, we were 'three sheets in the wind'.

Next morning we wandered down to the harbour where the nodding boats with folded oars looked like birds at rest. Nets festooned with bleached rags of seaweed were airing on the pier. A weather-beaten old salt was using his wooden peg leg to arrange drying dogfish. I felt as if I had somehow strayed into the past. He told me his leg had been taken by a shark, all very matter-of-fact, as if the fish had borrowed a book and forgotten to return it.

Emelio and I went diving in the shallows. The bottom was covered by a field of brown feathers swaying in the underwater breeze and boulders huddled together like bald men conferring. Some were toupéed with silky tentacled anemones and one dandy was wearing a big yellow sponge.

Although these untidy communities were less spectacular than the kelp forests back home, there was nothing to hide the view and I discovered many creatures that were new to me.

The sandy floor between the boulders was littered with the fat sausages of holothurians, relatives of starfish and urchins. Here, as all over the world, they were known as sea cucumbers, *pepinos del mar*. Emelio, however, called them *poy de burro*, a much ruder epithet they have acquired in a dozen languages. It describes their appearance and explains their reputation as an aphrodisiac. Without this firm promise surely no one would eat them. Sea cucumbers have some strange habits; if you handle them, they disgorge their viscera. The idea is perhaps to divert the predator with a snack while the cucumber sneaks away. It takes almost three months for its gut to regenerate and in the meantime it starves. Some sea cucumbers can breathe through their bottoms and have a tiny fish that lives in the backdraught. I don't know why and am reluctant to ask. Its name is *Proctophilus* (bum-loving).

Sea cucumbers were common enough here in the shallows, but five thousand feet down, lay the 'kingdom of the cucumbers'. There are 170,000 of them in every one of the billions of acres of the deep ocean floor. There are vastly more sea cucumbers on this planet than human beings.

It wasn't too difficult to dive into the sea from the shore, but getting out was a different matter. There were no modest waves here, only tumbling walls of water. To swim in towards the shore through the white fall of the waves was dangerous. I soon discovered that the trick was to turn and be carried in feet first by the surf. That way my fins took the brunt and with a bit of luck I might be flung ashore with the minimum of mangling and then grab on to the rocks before I was sucked back. The lava was puckered into needles and knives and spiked with black urchins. Gloves, it turned out, were essential, but I didn't have any.

It was safer to dive offshore from the dinghy. Emelio rowed and I sat on the tiny aft deck. The oars were so long that they crossed in front of him and to prevent them from colliding and clacking his knuckles, he had to keep one arm more extended than the other. It looked far more difficult than the method used by the rest of the rowing world.

The Atlantic was at its calmest, but even on still days the ocean is restless, as if being restrained against its will. It heaved great sighs of impatience and the boat rose and fell on its chest.

We dropped a tethered boulder to act as an anchor, put on masks, fins and snorkel and tumbled overboard into the chill Atlantic, leaving the boat to seesaw in the swell. I was dispatched to ensure that the boulder was firmly wedged on the bottom.

Emelio went straight down, his harpoon gun extended in front of him. He descended vertically on a wrasse and speared it just behind the head. The water was about forty feet deep, and down he went time after time and invariably returned with a fish. Emelio made it look simple but it wasn't. From above, many fish are as thin as a blade.

I had no taste or skill for spear fishing. I was a tourist taking in the underwater sights. Great rocky outcrops stood like Gothic cathedrals full of cloisters and dark corridors, and moray eels peered out from the entrances. They have such a surfeit of cranial cutlery that their mouth won't close. The saliva of morays is said to be toxic, but they needn't bite. They are so ugly they could just frighten you to death. With the ancient Romans they were a cult. No expense was spared to provide them with roomy ponds and long conduits to the sea to supply fresh sea water. The mother of Emperor Claudius hung gold rings from the snout of her favourite eel. They were kept as pets but also for dinner; Julius Caesar was reputed to have served up thousands for a single banquet. The Romans believed that in Hell they were employed to tear out the entrails of the wicked. No one doubted they could do it. When a party began to pall, the Romans flung a slave into the moray pond and watched the water change colour.

The sea floor was a labyrinth of tunnels. I found a larger cave and swam in. It was a stupid thing to do. I could probably hold my breath for a further ninety seconds or so and if my fins had stirred up the mud on the cave floor it might reduce the visibility to zero, and ninety seconds isn't very long to find the exit. In the darkness I saw something move. It was big, or so my imagination assured me. I decided I didn't want to find him and certainly didn't want him to find me.

I swam into a dense shoal of saupe, silver bream with golden stripes. It exploded like shrapnel then reformed in a swirl of silver. I stayed back and watched the school sway first one way then the other, like a flag unfurling in the breeze. It mesmerised me, just as it is assumed to do to predators. If there is safety in numbers, there is even greater safety in larger and more mobile numbers.

Although I could discern the individual fish, it was impossible not to see them all as a single organism whose parts, though separate, somehow stayed together in a flowing mass as if bound by a mysterious magnetism. No matter how sudden the change of direction, not a single fish was out of step. They swooped and fluttered and turned as one. It was magic.

I thought that perhaps a dominant fish led the pack, who fell in behind like a well-schooled chorus line, but it wasn't so. When the shoal turned, one side instantaneously became the front and led the flow. There was no leader, just a leading edge. Who then made the decisions and how were the instructions transmitted in an instant to all the others? It would be over thirty years before computer simulations showed that no overall supervision was required. Each individual can see and feel every twitch of those adjacent, and need only follow a set of simple, local rules telling it how to react to the behaviour of its immediate neighbours for the entire school to act in unison. It's still miraculous though.

Every day we went fishing for dinner, but never with a rod. Sometimes we dived, sometimes we lowered a hook baited with octopus meat and, gazing down through a glass-bottomed box, moved it towards the fish so that they couldn't miss it. Other times we cast a net and herded the fish underwater towards it.

Our victims were gutted, filleted, salted and dried in the sun. Every day we ate a dozen or so fried with maize or sweet potatoes and they were delicious. It is vital to dry fish quickly, for in warm, humid climates they are a magnet for maggots. Blowflies lay their malignant eggs in the gill cavities and eye sockets.

A far older method of fishing was still practised on the island. The most common local shrubs are *Euphorbias*, cactus lookalikes that contain a toxic milky latex. Centuries ago the *guanches* discovered that it could kill fish, but still leave them safe to eat. If a shoal swims into a suitably narrow inlet, the fishermen dam the entrance and pour in the latex. The fish float to the surface and are scooped up.

'Tonight,' said Emelio 'we go fishing for *calamari.*'

The dinghy felt very small even when close to shore, but far out to sea, it shrank to insignificance. The lights of the coast had stolen away and we were lost in the stars, little more than a wooden shoe alone on the sea. The mighty *Titanic* had foundered on an iceberg. We would be doomed if we collided with an ice cube discarded from a spent Martini. I kept my eyes peeled for a yacht manned by tipsy revellers.

The silky ocean was shot with moonlight and we were to jig for squid that darted in the darkness hundreds of feet below. They use a form of jet propulsion and are the most elusive and agile creatures in the sea. Towed nets rarely catch them for even in the deep they can see the fizz of green luminescence approaching long before they are ensnared.

Our method was more devious. We lowered down a metal plummet, slender as a sardine. It was topped with a rosette of hooks like a crown of thorns. We paid out over three hundred feet of line and then began to jig. We jerked it by hand and the boat's regular rise and fall added a deeper rhythm. Far below our silver lure twitched and slid up and down, and in the gloom even the wily squid were fooled. They pounced on what they thought was a fish and their tentacles became ensnared on the hooks. When Emelio felt the line become heavy he hauled it up and every time it terminated in an embrace of squid.

In the torchlight I saw them blushing with wave after wave of iridescent colour. Were these neon ripples a symptom of distress or a warning to other squid that there was danger here? The squid's skin has pigmented bodies in red, yellow, brown and black that it can expand and contract at will, plus tiny 'mirrors' of silver and gold, so it can change colour and transparency in an instant. But squid don't flash at random – they signal. There are specific displays to attract mates or repel rivals, to baffle prey or frighten predators.

There is no sensation that remotely resembles being adrift on a windless ocean at night. Our boat rode the moving hills of water, rising and falling with the easy motion of the swell. From a valley where walls of water were head-high all around us, we were borne upwards. Then, on the crest, just a moment's hesitation before the long slide down again into the next dark trough. But the unnerving thing was that all this movement took place in silence.

We had been at sea for almost two hours. Dense clouds obscured the moon and I had no idea in which direction the shore lay. Emelio licked his finger and held it up to feel for the wind. 'That way,' he said. We rowed into the night for over an hour and at last I was relieved to hear the growl of the surf somewhere ahead of us in the darkness.

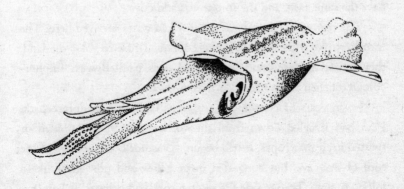

The
Turquoise Pool

Jameos del Agua, Lanzarote

The lava flows lying inland from Punta Mujeres receive less rain than
the Sahara Desert. There are no dry ravines for there is nothing to
carve them, nor anything soft enough to erode.

For some reason that now escapes me I decided to camp in the
middle of this arid wasteland. There was no way I could pitch a tent
so I lived in a small cave that had been partly walled up, probably to
corral goats, judging from the droppings on the floor. I found that I
had forgotten to bring fuel for my stove so for a week or so I cooked
over an open fire fuelled by goat dung. Soon everything seemed to
have the same taste and the smoke was addictive.

The region was a wilderness. How had goats survived here? The
lizards awoke with the sun, but as soon as it rose to roast the land,
they retired underground. Every rock was painted with sulphur-
coloured lichens with names such as *vulcanea*.

Seven years of continuous eruptions had left a sculpture park.
Lava had treacled down from the volcano, but then wrinkled or
twisted itself into ropes. It had begun to cascade over the edge of the
roof of my cave, but congealed in mid-flow and now hung like a
lolling tongue. To one side the molten rock had dripped as wonderfully as candle grease. The ground was littered with 'bombs' bigger

than rugby balls that had been hurled from the crater several miles away. Elsewhere the black volcanic glass had been drawn up into needles. By noon, everything was too hot to touch.

Yet someone had once lived nearby, for I came across an ancient stone structure eight feet long that looked just like a pathologist's dissecting table. I was told later that it was a sacrificial altar used by the *guanches*. Well, maybe. The *guanches* had strict codes of behaviour, but were brutal. Here on Lanzarote, murderers were executed by resting their head on a flat stone and then bashing out their brains with a rock.

There were no roads or tracks here and I had few visitors, but one day two lads turned up with a weasel. They were hunting for rabbits and gave me one for tea. It was the thinnest animal I had ever seen and I wasn't over-adept at fleshing still-warm animals. There may be more than one way to skin a cat, but I was assured there was only one way to undress a rabbit. One of the lads made a small incision behind the neck. He placed his lips to the wound and gave a mighty blow. The rabbit inflated and then he simply peeled off its skin as if removing a rubber glove. He snipped off the feet and nose and handed me a naked rabbit. Tough but tasty.

The sparse, ephemeral vegetation had shrivelled to straw and blown away. It was ideal cactus country, but they are inventions of the New World and there are no cacti here. Instead, *Euphorbia*s have stepped into the breach, not the leafy spurges of our dunes, or the scarlet-bracted Christmas poinsettias, but bush-sized candelabras. Leafless succulents do best when the hot winds blow in from Africa. On the larger islands there is the impressive *Euphorbia canariensis,* which forms an immense thicket of spiky towers. Pliny tells of the first ever expedition to the Canaries, sent by King Juba II who ruled the Roman protectorate of Mauritania around the time of the birth of Christ. The expedition reports described *Euphorbia canariensis* for the first time and botanists later named them after the king's physician, Euphorbus.

Euphorbia canariensis is so cactus-like that even an Arizonan wouldn't notice the difference. All these cactus mimics are endemic to the Canaries; they are found nowhere else on earth, although elsewhere other groups play at being cacti when there are no real ones around.

Islands are wondrous places and hotbeds of idiosyncratic evolution. Lanzarote and Fuerteventura were probably once fragments of Africa that became separated from the mainland long ago, but the other islands in the Canaries are volcanoes that arose fresh from the oven beneath the ocean. They were once barren, but were progressively colonised by whatever blew in on the wind or came ashore on the ocean currents. Some would-be residents hitched a ride on driftwood or volcanic pumice. Over time the squatters evolved into new forms different from those that had arrived. Oceanic islands have gone their own evolutionary way. The one thing they have in common is that each is unique.

Islands are also often devoid of the predators and competitors that have driven evolution on the continents, so the pressures are different and many evolutionary experiments that would have failed on the mainland may succeed here.

Whichever species become established eventually occupy the niches left vacant by those that don't. They give rise to new forms suited to new roles, for ecological imperatives ensure that closely related species find it difficult to co-exist unless they exploit different niches. Being different reduces competition and increases the chances of survival.

Over a fifth of the animals and almost a third of the plants on the Canary Islands are endemic, but not all of them have evolved anew here. Islands are also lifeboats in an uncertain world, havens where species can survive when they are becoming extinct elsewhere.

Many species now confined to the Canaries were once wide-spread. They are the remains of a lost flora that thrived 15 millions of years ago. The Canary dragon tree formerly ranged across the globe and its closest relatives are now found half a world away on the other side of Africa and in the East Indies; a forest of Canary pines once stretched to the Himalayas, where its nearest relative still survives. These are the survivors of ancient catastrophes and climatic changes. Islands are where threatened species hide.

Everywhere there were tubes and tunnels where the surface of lava flows had cooled and crusted over, insulating the hot lava within and allowing it to flow on much faster than before. Often it ran right through leaving an empty tube behind as an echoing memory of a resting volcano. Sometimes there was a small tube above with a larger tunnel beneath. Some tubes were just little stone burrows in which shearwaters nested. I could hear the chicks chuntering inside as if they were on the telephone to a friend. Other tubes were big enough to let through a train if it strayed from the London Underground.

I have no head for heights but I was a keen caver. I thought nothing of dangling over a hundred-foot pitch in darkness, because the drop was invisible. Although I have always loved being under-ground, I'm not sure I want to be buried. I suspect the novelty wears off.

Lanzarote has the longest lava tunnel in the world. It runs four and a half miles to the coast from the La Corona volcano and continues under the sea for a mile or so more. Where the landward stretch dips into the sea, the end of the tunnel is blocked by debris and in this cul-de-sac cave is a stunning underground pool called Jameos del Agua. It was wreathed in purple gloom, but the sun entered through two large holes in the roof where long ago a belch of gas popped out a circle of crust, as neat as a cork from a champagne bottle. Slanting columns of sunlight illuminated a turquoise lake eighty feet long. It took my breath away. No stage set could have been more dramatically lit.

The sound of the surf could scarcely be heard inside the cave, but its presence was felt, for the level of the pool rose and fell with the tide as the sea water seeped in and out from the ocean four hundred yards away.

I stripped off and walked into the water with reverence. The bottom was a tumble of boulders. From a thousand crevices long glaucous ribbons protruded. When I tried to touch them, they shot back into their hole with the sudden recoil of stretched elastic. What on earth were they?

I sneaked up on one and chopped with my diving knife before it vanished into its lair. The detached ribbon stayed alive for hours rhythmically extending and contracting. It was soft and frilled and had a groove all down one edge, and the tip was forked. It was distinctive, but was it a worm or what? As I stared, it contracted into a question mark.

I was not the first person to be puzzled by this creature. Long before, a biologist had examined just such a piece and described it as a new species. What neither of us knew was that there was another bit. Hidden in the crevice sat the fat green plum of its body. The thin gelatinous concertina that I had captured was just its proboscis searching for food. It is called *Bonellia* and belongs to an obscure group I had never heard of.

The 'plum' is the female and she is seventy times bigger than the male, a tiny parasite that lives quietly among her folds. The larva is of undecided sexuality. If it encounters a female, it becomes a male and slips into the groove in her proboscis. In the absence of female company the larva simply becomes a female and then, like most females, waits for ever for a male to call and when eventually he does, he gets up her nose.

There were also some tiny squat lobsters, but I didn't pay much attention until I came back that night for a dive in the dark. The boulders were now swarming with them. They were bone white and so were their eyes. Why should they be blind in a cave that was lit through holes in the roof?

They thrived here for they had no enemies in the cave and plenty of food as the light encouraged tiny algae to bloom in the water. I hung in the water for a long time watching them trundle around, then they decided that the show was over and returned home. And so did I.

Their closest relatives are found in the deep ocean and most of them also have pale bodies and feeble sight, but not all. One species found recently beside deep-sea vents has goggle eyes. It was thought to have become extinct in the Triassic. The deep ocean has a longer memory than the land.

How was it that these little lobsters now inhabited two such distant places, the great abyss of the ocean and a little cave on Lanzarote? As with the dragon trees, curiously disjunct patterns of distribution are not unusual. The world had been a troubled place throughout the evolution of living things. Many groups were once more widespread than at present but the land rose and fell, the sea came and went and its temperature fluctuated. Those species that survived were often just the lucky few that happened to be in a place buffered from the perturbations raging elsewhere. If I were more aquatic and had to find a safe place if the worst happened, I would choose to hide away in the ocean deeps or underground where conditions might stay constant.

I met Señor Soto who had been put in charge of 'civilising' the cave. He was planning to paint it with light and pave it with octagonal basalt sets. 'What do you think of this?' he asked, showing me the plans.

'I'm sure the tourists will like it,' I said.

Tourists get everywhere and the little squat lobsters in Jameos del Agua have suffered deteriorating conditions in the cave ever since the pool was opened for visitors. Now the numbers of lobsters have crashed and the males are only half as fertile as they once were. It would be sad if they had found shelter from some of the greatest cataclysms in Earth's history only to be eliminated by a gaggle of trippers.

The big tunnel now echoes with disco music. The cave of the turquoise pool has been illuminated by coloured spotlights. One of the most stunning and eerie places in the world has now been given an atmospheric makeover and its atmosphere has been lost for ever.

Lanzarote had been chosen for our expedition because it was unspoilt, and almost undiscovered by tourists, but we were only just in time. Then there was only one hotel on the island and a guidebook warned the visitor not to expect any social life or organised amusement. Today its economy is dominated by tourism. Now, there are over a million visitors a year to the island. In 1963, most of the young people on Lanzarote had their bags packed, waiting for the day when they could leave for Tenerife or mainland Spain to seek their fortune. Now 75 per cent of all males under thirty are employed in the service of tourists. The black beaches have been exchanged for golden sands fortuitously blown here from the Sahara on the sirocco winds, or so the brochures claim. In the arid wilderness there are now five-star hotels awash with swimming pools, fountains and fake waterfalls.

It is curious that we flock to the crowded places to relax and get away from it all. Nowadays, wherever there is surf there's sewage. We often spend our holiday imprisoned in the hotel pool so as to avoid the suspect sea. We don't get stung by jellyfish, but by the local tradesmen. At night, even the most distant seaweed sways to a disco beat, and turtles head in the wrong direction to lay their eggs because at night the shoreline is brighter than the moon. Exotic shorelines are deforested to be replaced by specimen palms planted on the hotel lawn.

In exploration, timing is everything. Yesterday's expedition is today's excursion and tomorrow's packaged tour. The word 'remote' has lost its meaning and adventure is sought in ever more pointless stunts. The daring bustle to be first to balloon around the world backwards or cycle across Antarctica, and we celebrate their achievements, for we are too polite to say, 'Haven't you got anything better to do?' Even the explorers are unsure why they went, and the old excuse, 'Because it is there', is perhaps the best answer to be expected from brains starved of oxygen from too much bivouacking at high altitude.

There are only two great wildernesses left: outer space and the deep ocean. Space is a dusty void full of the wreckage of defunct satellites; rockets have to dodge their way past en route to Jupiter. We justify the expense of escaping from gravity by claiming we are heading for the stars, but in truth we are only peering at lifeless planets.

Only 1 per cent of the ocean has been explored and its dark floor twinkles with just as many stars as the heavens, but these are the luminescent winks of unknown creatures of wondrous design and mysterious behaviour. You can keep the dead expanse above our heads, given a choice of which stars to explore, I would go down to the living illuminations beneath the sea. Who knows what aliens await discovery?

Voyagers

Northern coast of Spain

I said goodbye to Emelio and his family and returned to Arrecife to help prepare for our departure. I distributed our surplus stores to the locals, but goodness knows what they made of sixty-four packets of Scots Porridge Oats and an entire crate of Grape Nuts. They had no need for cavity-wall insulation.

Something smelled bad, very bad. It was traced to an unopened box that wasn't ours. Presumably it had been picked up by mistake on the dock seven weeks before. Immersed in a mass of suppurating food were half a dozen corsets. Clearly, somewhere on the island was a desperate woman whose body was completely out of control. It seems apt that whales that wear their blubber with such elegance should provide the struts for corsets to streamline the chubby folds of humans.

We made our farewells to Don Mariano and set sail for Santa Cruz de Tenerife. I fell in with an American who had been provost Marshal of Wigan during the war. He had imprisoned Gracie Fields for undermining the morale of his troops – and who could blame him?

In the local church was a tattered flag left behind by Admiral Nelson, together with his arm, when he landed here in 1797. Nelson's fate was to be progressively disassembled wherever he went. He had recently lost an eye in Corsica and many of his teeth were missing. From what was left a hero was fashioned.

He had been only thirty-eight years old when a musket ball severed the artery in his arm as he leapt ashore at Santa Cruz. The ship's surgeon removed the arm and Nelson's only complaint was that the trimming knife was uncomfortably cold. He later issued an order that all the surgeons' cutting tools should be warmed before use.

It is amazing that this remainder of a man, whose surviving teeth were so discoloured that he only 'smiled' with his mouth clamped shut, could have attracted such a renowned beauty as Lady Hamilton. Perhaps it was the power of celebrity that caused her to long for his arm around her.

Our last meal in a local hotel was, according to the English menu, 'Glossy balls of heifer'. As they were neither shiny nor spherical, I have no idea what they were, and perhaps it's just as well.

The *Begona* cast off at midnight and eased slowly away, with a feeling for the solemnity of separations. Ludicrous martial music sounded over the Tannoy, yet it was strangely affecting. Waving friends and relatives, all strangers to me, lined the dock as we parted from the land. Everyone was in tears.

'*Estancada*,' said Martin.

'Yes,' I agreed. I knew what he meant.

The ship called in at La Coruña. A line of women stood on the dock. They were all in their late fifties and dressed in black, some with loosely knitted shawls like Victorian fishwives. To my astonishment they heaved luggage and crates on their handcarts and then shot off at the trot down the street. They were the harbour porters and woe betide any pedestrian who thought the pavement was intended for them.

We strolled through a maze of narrow streets and houses with discarded stucco, unhinged shutters and crumbling once-decorative scrolls. Provincial Spain was falling down a flake at a time. Platoons of disabled beggars and ragged children rested beneath the perfumed trees, counting their day's takings.

The bar where we ended up was lit with a dim yellow light to lend an oriental enchantment to the prostitutes at the counter. One was bone thin. She had a sad face and a deep scar that ran the length of her cheek. Probably she had to take the rough customers with the smooth. The other was voluptuous de luxe with breasts like party balloons. She swayed to the rantings of a jukebox and made jokes at the expense of the fresh-faced English lads, daring us to put some money in the slot. She didn't know that we were explorers of more exciting terrain than hers. I had kissed a moray eel and embraced an octopus. What more could she possibly have to offer?

With the help of cognac, eddies of conversation blew up into gales of laughter. The barman asked us to keep it down a little. 'You'll have to forgive my friends,' said Martin. 'They have been at sea for a very long time.'

'How long?' said the plump tart.

'Well,' said Martin as if totting up the months. 'Almost two days.'

I had become very fond of Martin, but that was luck. Expeditions are testing times because people are flung together in difficult conditions. Getting the right personnel is vital, but you rarely do.

Only nature achieves perfect teamwork. From the deck of the *Begona* I had spotted a Portuguese man-of-war, one of Haeckel's siphonophores. It had a bluish float shaped like a Cornish pasty and below, flowing out with the current were trailing streamers armed with toxic barbs. But the most interesting feature of this creature is that it is not *one* creature at all, but a colony, an intimate amalgam of several interdependent animals. The float is one entity that buoys up all the others; the streamers are animals up to fifty feet long that paralyse the prey, the feeding polyps consume it and share out the proceeds among all the others in the kibbutz, including the reproductive polyps that propagate on behalf of the entire team. If only all expeditions could be so well organised, with someone reliable to arrange the transportation, another to supply provisions and someone else to protect the team from outside interference, leaving me free to take care of reproduction and associated matters.

Every expedition should be run twice, once to get it wrong, and again to get it right. Unfortunately, most explorers only experience the first trip.

Nonetheless, we had been on an adventure. True, we were probably at greater risk when dodging the rush-hour traffic on Lime Street, but it didn't feel the same. I had gone to immense trouble to allow the Atlantic Ocean to fling me against snarling rocks, had spent days in a cave in a blistering desert, eating anorexic rabbit and inhaling goat-dung fumes until I had hallucinations.

I had also risked all manner of fatal infections by repeatedly stabbing my finger to persuade timid fishermen that it didn't hurt and they should give a drop of blood to the medics. My reward for all this blood letting was hepatitis on my return.

Martin came to visit me, for, having shared our small adventure, we remained friends. He was wonderful company and possessed great charm. There was a twinkle in his eye that I thought could never be extinguished, but I was wrong.

He went on sabbatical to the United States and took his family with him. They were in a pile-up on the freeway. His wife and two of their children were unharmed, but Martin and another child were killed outright. Their fourth child was flung into a coma from which he never emerged. He died thirty years later.

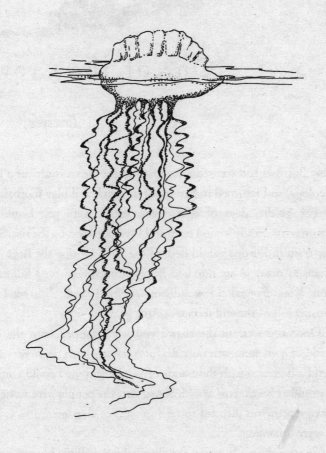

One of the Crowd

Liverpool

Bunny Burrows had secured a studentship for me to study for a PhD in ecology, so I returned to Liverpool to watch and play football.

The golden days of Merseyside soccer were just beginning. Everton were already a good team and Liverpool was on the rise. Soon every team in Europe would hear of the Kop. During the Boer War perhaps as many as six hundred British troops died on a hill called Spion Kop. Several Liverpudlians survived and christened the imposing hill of the end terrace at Anfield.

Once you were in the crowd you were trapped until the final whistle. There were several rules of survival, such as never stand behind a barrier – with thousands of people surging behind you, the bar wouldn't break, you would. Sometimes the people were so tightly packed that it was difficult to take in air. For a moment it was as if you were drowning.

Another danger was that the chap behind, having downed twelve pints, would rather wet your pants than his. Even so, the Kop was a wondrous place. For an evening match in December I got there early to ensure a place. It was snowing and the ground was eerily lit by floodlights on at half-power. From the side stand, the Kop was an immense void, darker than the sodium sky and a-twinkle with far more stars as the crowd dragged on their ciggies. Smoke crept out into the night from under the lip of the roof like an inverted waterfall.

People who fainted were hauled aloft and passed down hand over hand above the heads of the crowd. From a distance, the stiff bodies seemed to glide down the steep slope as if on their final journey in a majestic crematorium. Veteran fans often asked to have their ashes scattered there.

I also played football, but with difficulty. Bunny endeavoured to keep me in working order by forbidding contact sports. It was rumoured she was seen crossing my name off the team sheet.

Matters came to a head when I sustained a black eye in a collision on the pitch. For three weeks, as the bruise slowly yellowed, I managed to keep my left side turned away from her whenever she came into the room. Until one day I was facing the wrong way.

'How did you do that?' she asked.

'I walked into a door,' I said.

She didn't believe me, of course.

Shortly afterwards, on an excursion to the Isle of Man, one of the students got friendly with a local girl, having rescued her from a drunken fisherman on the dance floor. When we left, the entire street was full of fishermen waiting for us. We fought them intermittently all the way back to our hotel. Next morning Bunny immediately spotted that my black eye had changed eyes.

'Another door?' she snapped.

'No,' I confessed, 'I accidentally walked into a fisherman.'

Albert Camus said that playing football taught him all he knew about life. But university teams like the one he played for were composed exclusively of educated fellows with future prospects. I learned about life when I turned out on Sundays for a team called Star United. It was mostly local lads with a sprinkling of university mates. Our number included weedy Bob Burley and the sluggish Ron Speed.

We invariably lost or won by the same score: six–two. One Sunday we even drew a game six–two. Because of an accumulation of fixtures, we agreed that each half would be extended to sixty minutes and count as a separate game. We lost one and won the other – both by three goals to one.

Burley had an undecipherable accent, but when felled by a late tackle, he leapt to his feet, whirled around and said, 'By Jove, Sir Percy, you've gone too far this time.' and ran the ruffian through with an imaginary rapier.

My pal Dave disputed the referee's decision and the official asked, 'Who's refereeing this match, you or me?' Dave unwisely replied, 'As far as I can see, neither bloody one of us!' He was sent off.

Every week the local paper carried a review of our performance: SHOOTING STAR GAIN REVENGE, and my favourite, HE SCORES WINNER THEN BREAKS HIS LEG. Our striker toppled from a heavy tackle and everyone on the pitch heard his leg snap. He took it like a man, for there was far greater pain to come. All he said, over and over again, was 'Bloody hell, the wife'll kill me'.

When our goalkeeper started going out (and staying in) with a blonde raver called Jackie, he became listless and lost interest in balls that were above shoulder height. He dived only in the sense that boxers take one in the fifth. If the game drifted to the opponent's end of the pitch, he leaned against the post and fell asleep. We replaced him with a fellow biologist, Steve Goodband, and the headlines for the next match was GOODBAND IS THE HERO.

Unemployment on Merseyside was high and for the local lads the prospects for employment were dismal. It would be so different for the students. Two of us became university lecturers in Britain and Dave went off to be the director of the Botanic Gardens on Gran Canaria.

Steve got an enviable job at the University of the West Indies. He was a keen yachtsman and took a non-sailing friend out in his dinghy to watch a race. The bailer fell into the sea and Steve, a strong swimmer, jumped overboard to retrieve it. It was the reaction save of a goalkeeper. The boat sped away and left him alone in the ocean.

The open ocean is an imposing place during the day, but alone at night surrounded by sinister phosphorescence and shadowed, at least in the imagination, by sharks, it must have been a place of terror.

His body was washed ashore the next day.

There is a release when you enter the water. You are immediately weightless and unshackled from everyday life. To swim naked is to be truly embraced by water; to feel it easing past your body is to become fluid.

Drowning is the ultimate release from a dependency on air. It is what water does best and it is wise not to forget it. Every year seventy bodies are hauled from the River Thames alone. There are legions that have sunk into the sea never to surface again. It used to be believed that the great weight of water in the oceans could compress that deeper down, increasing its density so much that any object would only sink until it encountered water as dense as itself. Thus drowned bodies would never reach the bottom, but remain suspended for ever at their prescribed depth. If this had been true, would we now have tourist trips to the interesting, everlasting graveyard in the deep?

Every week a ship sinks, often without emitting a Mayday. When ships founder, great crowds are simultaneously swallowed by the impassive ocean: *Titanic*, 1912 – 1,495 lost; *Lusitania*, 1915, torpedoed and sank in eighteen minutes – 1,195 lost; HMS *Hood*, 1941, broke in half within six minutes of the *Bismark* opening fire – three saved, 1,412 lost; *Bismark* sunk three days after destroying the *Hood* – fifteen survived, more than two thousand lost. Once a ship goes down, the blind sea closes over and only flotsam remains – until that terrible day when, we are assured, the sea shall give up its dead.

There are also the lonely deaths, both accidental and intended: lovely Natalie Wood who was careless and fell overboard, and Virginia Woolf who took great care to fill her pockets full of stones beforehand. There are some victims I have known, only slightly, but enough to feel the sadness: Per, the Scandinavian biologist for whom too many little things went wrong simultaneously; Mia, who succumbed to narcosis deep down; and Keith. He was destined to become Britain's leading underwater archaeologist, but when picking up artefacts in the shallows he leaned over too far and turned turtle. His weights fell towards his neck, air rushed into the boots of his suit and he couldn't right himself. He drowned upside down in less than a metre and a half of water.

And then, of course, there was Steve. Whenever there is an incident underwater and the ocean reminds me that I have no right to be there and, if careless, might be invited to stay, I think of Steve.

Caribbean Nights

Liverpool

Research students are apprenticed to one or more supervisors. Nowadays, supervisors see their students almost every day, but it was very different in the 1960s. Bright students were thought to thrive on freedom. I rarely saw Bunny and although she periodically fretted about this, after each encounter she forgot about me for months. This arrangement suited me fine.

A colleague told me that he began his research on an aspect of co-evolution (where two ecologically dependent organisms, like insects and the flowers they pollinate, mutually adapt to each other over time). After a few months it became obvious that the project was a non-starter so he changed to something different. Three years later he successfully submitted his thesis and secured his first job thanks to a glowing reference from his supervisor. His new head of department greeted him with 'Good to have you on board. Just what we need, an expert in co-evolution.'

My friend suddenly felt sick. 'Actually, I know rather little about it,' he confessed.

'No need to be modest, old man, your supervisor made it clear you were just about the world authority on the subject.' Despite getting the job on false pretences he went on to become a Fellow of the Royal Society – but no thanks to the benefits of supervision.

Today's students have to expend far more energy avoiding their supervisors when they haven't completed an assignment.

The Liverpool campus was expanding. The site had been cleared for development and had become a wasteland of pubs. Imagine a grid of streets with all the houses gone, but the pubs still standing at the corner of each vanished block.

On the periphery of the precinct the old shops remained. Almost all of them sold everything in general and nothing in particular. Window gazing was a delight:

> Ladies panties – all colours
> women's sizes only

Above his door a Chinese tailor had written:

> Bespoke garments
> Ladies given fits upstairs

At night, music emerged from clubs with exotic names such as the Jacaranda. The owner, Allan, was the first manager of the then Silver Beatles. He sent them to Hamburg, got them their first booking at the Cavern, then carelessly lost them to Brian Epstein.

Some clubs were more difficult to get into. Often, they were former shops that had their windows entirely painted black. The door had a small flap like that in a cuckoo clock. Surely doors with a peephole had something to hide, so I knocked. The flap jerked aside and the hole was occluded by a big black face: 'Wadda ya want?'

'May I gain admission, please?'

'You a member?'

'Well, not exactly.'

'Then bugger off.'

The rhythms of Jamaican ska seeped out into the night, but this dark and mysterious world remained a secret to me, and perhaps it was just as well.

I sometimes walked to the cinema with my mates along the once grand Georgian terrace of Upper Parliament Street, which had fallen into disrepair and disrepute. Beautiful West Indian girls hung from every upstairs window. 'Hey deh, boys, wanna come up for a thruppenny one?'

Although they had accurately assessed our means, even in those days you didn't get much for threepence. But the women were so luscious it would have been worth it, if it hadn't been for the big pimp hiding in the wardrobe ready to leap out wafting a razor.

Outside the chemistry department was a billboard for the Pavilion theatre. Every week was worst than the last: *Naughty but Nice*, *Naked As Nature Intended* and *Strip, Strip Hooray*. We couldn't resist *Nudes, Nylons and Nonsense* and became avid theatregoers.

The star was not sexy, but was almost certainly a sexagenarian. She wore a bruised peach body stocking and hid behind feather fans plucked from the world's biggest ostrich. At intervals there were tableaux of unbridled cellulite. These Aphrodites in Anaglypta were forbidden to move. One sneeze and the constabulary would have rushed onstage, for there was invariably an off-duty copper in the audience.

One evening Dai, my roommate, and I went to a pub afterwards. It was all stained glass and stained carpets and we were accosted by an ancient tart with a heart of gilt. 'Give you a good time, sonny?' she said to Dai. 'It's only a fiver, half-price for pensioners and schoolboys.'

He flushed vermilion and said 'I haven't got a fiver.'

'Never mind,' she said. 'I'll lend you one.'

Her mates roared with laughter.

'I wouldn't dream of . . . of . . . with you,' said Dai sniffily.

At this she thrust her hand into her bag. I thought she was going to pull out a Smith & Wesson. Instead, she drenched him in Evening in Paris and rasped, 'Explain that to your mam you pompous little ponce.'

This was Liverpool in the sixties, the centre of the known universe. People have often asked me what it was like to be there. If only I'd known it was going to become 'Liverpool in the sixties' I'd have paid more attention.

We were supposed to have spent the entire decade zonked out of our minds, but I was too busy having a good time. Bunny had taken on a new graduate student called Win Price. She was obviously bright and I would like to say that I was first attracted by her intellect, but in the age of miniskirts there was so much to admire that I got distracted.

We became friends long before we became lovers, quite the opposite of my usual practice. Our first date was after a departmental party with plenty of sherry and wine. Fortunately I was hardly affected. The building, on the other hand, was in a terrible state. The walls began to lean at a terrifying angle, the doors oscillated as I approached and the stairs began to concertina if you attempted to descend. There is nothing more frightening that a building that can't take its drink.

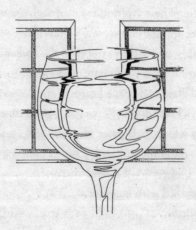

He Who
Dares, Swims

Port Erin, Isle of Man

Somehow, despite these distractions, I had to do some research for my
PhD. Since my plants grew underwater I would have to go down to
meet them. To get properly trained I joined the British Sub-Aqua
Club.

The public baths at Birkenhead was a tired Victorian emporium
with acres of cracked and crazed tiles. Every Thursday night it was
booked for dozens of cracked and crazed divers. I have always hated
the echoing void of swimming pools, the aroma of someone else's feet,
verrucas donated by generous strangers and enough chlorine in the
water to gas a regiment.

I was an accomplished sinker. Unfortunately, all the preliminary
tests involved remaining *above* the surface. I had to swim two lengths
of the pool freestyle, then backstroke and then don a diver's weight
belt and swim one more. This last was an ordeal. I rapidly lost forward
momentum and the far end of the pool seemed to retreat at about the
same rate as I advanced. Slowly, like some doomed liner, I sank
inexorably below the water and never reappeared; there was no way I
could regain the surface. Ten seconds before drowning I reached the
finish and sucked in so much oxygen in one breath that everyone
present was in danger of suffocation. 'That was impressive,' said the

instructor who thought I had been showing off by swimming the last length underwater.

The next test was life-saving, for which I had to tow a passive victim the length of the pool then give him artificial respiration. My victim was enormous, six feet four and as broad as a Scouse accent. Nonetheless, I managed to haul him down the pool then discovered that I was expected to resuscitate him on land. So, first, I had to get him out of the water. I weighed 150 pounds, he was 203. It wouldn't have been so bad had I towed him to the shallow end so that I could have stood on the bottom and thrust him out on to the side. At the deep end, every time I pushed him up all that happened was that I sank. By the time I 'rescued' him, I was the one in need of resuscitation.

After having dried and changed I went to see the instructor. Miraculously, I had passed. As he signed my logbook, he said, without a trace of irony, 'I've never seen anyone do it from the deep end before.'

I smiled modestly, as if to say, 'What other end is there?' When I nonchalantly leaned forward to sign the document a cupful of chlorinated water flushed from my nose and swamped the table.

I was soon qualified to roam the underwater plains and rustle sea horses. From now on, as befits a scientist, I would dive in metres not feet. Unfortunately, I was in Liverpool and the plants I was to study were growing on the Isle of Man. 'All right,' said my prof, 'the department will pay your air fare for a trip to the Isle of Man every month.'

I would work at the Port Erin Marine Biological Station, Professor Herdman's old hideaway on the island. An ex-research fellow from the marine station warned me what to expect. 'First time I went to sea with the prof, he got me to hold his false teeth while he was being seasick.'

I knew little about the Isle of Man. According to the Victorian botanist Mrs Gatty, it was an 'unusually charming island, with its soft climate, bad farming, bare-legged begging population, beautiful scenery, remarkable antiques, and last, not least in love, its exquisite sea-weed shores'.

For two years I spent several days a month in Port Erin at Mrs C's boarding house on the seafront. I recall lumpy meals, lots of tea and endless confusion about the name, Marine Biological *Station*. 'I'm just off to the station,' I told Mrs C.

'You're leaving already?' she cried. 'Was it the lumpy lunch?'

Even the railway station was confused. A sign outside declared:

> The 4 o'clock train will be a bus.
> It will depart at 20 minutes to 5.

Accommodation in summer was hard to find. A researcher went off on a short sampling cruise and came back to find the landlady in his bed – she had rented out hers to visitors. I was befriended by Peter and his wife Carol. They owned one of the big hotels on the seafront, and I was invited to stay with them out of season. In the wilds of winter we occupied a mothballed hotel with ninety empty bedrooms.

Peter, modest and quiet, was the engine beneath the bonnet of the hotel, whereas glamorous Carol manned the front office and dealt with the customers. Years later I recognised the woman decorating the front page of the *News of the World*. It was Carol starring in an article entitled NAKED ROMPS IN WITCHES' COVEN. Apparently she had become the high priestess of a group whose devotions involved casting spells and prancing around in the nude. Imagine my surprise, and disappointment, at not having been invited to join the congregation. I would have leapt at the chance to become a wizard.

At the marine station, coffee was served in the library at eleven and most of the staff managed to get there in time for it. The table was strewn with copies of *Punch* and *Scottish Field* and sometimes a lively tug-of-war developed to secure the latest issue. The director once said to a colleague, 'What are you doing in the lab on a beautiful day like this? You should be out on the golf course.'

The local girls were desperate for new faces and understandably interested in the young men from the marine station. Early in the conversation they would ask casually, 'Are you married?' Then, 'How much do you earn?' Quite often the conversation ended right there.

My first research sea dive was in February with two staff from the lab, Norman Jones and Joanna Kain. There was no exploring or taking in the underwater scenery. This was my new workplace where you went down and got on with the job.

When we emerged it was snowing, yet we stood waist-deep in the frigid shallows discussing the merits of a new snorkel. The snorkel was just a plain bent tube with a mouthpiece at one end, yet the debate continued interminably while I sank quietly into hypothermia.

Joanna was a New Zealander who had come to Port Erin in 1956 and was intrigued by the seaweed forests. She signed up at Captain Hampton's diving school at Dartmouth in Devon. It was the first aqualung training course in Britain. For five pounds he promised to get you to the 'naval standard of frogging' in a single day. Depending on your gender and age, you qualified as a 'manfish', 'mermaid' or 'tadpole'. His menfish included Arthur C. Clarke and David Attenborough.

After a ten-minute briefing in the boat, Hampton sent Jo down the ladder to the sea bottom where she would then take out her mouthpiece as a confidence-building exercise. Six metres down, the mouthpiece leapt from her hand, shot above her head and gushed air into the water. Hampton realised that none of this was getting into Jo, so he hauled her up. If she had held her breath, the expanding air inside her lungs could have done immense damage. Fortunately, Jo had read this on the train down to Devon, and exhaled like mad.

Reg Vallintine, another diver trained by Hampton, was sent down alone without fins to 'walk around a bit on the bottom'. Reg was immediately entangled in head-high seaweed. Returning to the surface prematurely, he explained the problem.

'No, no,' countered Hampton. 'It's just sand.'

'But I've just been there. It's thick with tall weed.'

'Just imagination, old boy.'

Hampton had been a test pilot and when asked if he could fly a Spitfire, replied, 'I think so,' and climbed into the cockpit. After making a high-speed pass upside down, he landed not on the runway, but on a side road.

'What went wrong?' asked an anxious artificer.

'Nothing,' Hampton replied.

Jo soon became one of the first female First Class divers in Britain. She wrote the chapters on scientific diving in the early manuals of the British Sub-Aqua Club and helped to draft the first Code of Practice for scientific divers.

Jo was immensely practical. If you blew a tyre, odds on she could produce a wrench from her bag and change the wheel. If you had an accident, she could be relied upon to staunch the blood or perform a tracheotomy with a Biro.

She also trained many people to dive, including several of Bunny's students. We were known as Jo's Boys. On each visit to Port Erin I managed several dives. Mostly I went down alone on a safety line, with the skipper, Johnny, guiding the dive boat above. His mate Louis held my line, and every few minutes jerked out the code for 'Are you OK?' and I would tug back 'Yes'. Sometimes his jerks were sufficiently violent to displace me and uproot my plant. I hope Louis wasn't listening to the bubbles. Diving alone was never a problem. The main danger was if you stuck your hand in a hole occupied by a conger or a lobster. I know a diving biologist who surfaced and handed his attendant a lobster with one hand and the end of his finger with the other.

Louis and Johnny were salty encyclopaedias of folklore, mariners' tales and good old gossip. Long before anyone else, they knew that Norman and Jo were more than just dive buddies for 'their bubbles were far too close together'.

Jo also had a brief affair with Donald. Donald was a bottlenose dolphin who adopted the Isle of Man for three years from January 1972, and was a likely companion whenever you dived around the south of the island. Once in turbid water he repeatedly torpedoed towards Jo, appearing suddenly out of the murk and swerving aside at the very last moment. Even the wash was enough to push her over. In a quieter mood he would just give her a nudge and tweak her flippers, or peep over her shoulder to see what she was doing. One day she decided to stare at him with her mask almost touching his eye and he rushed around her leaping and turning somersaults.

Then one day Donald became much bolder. He sneaked up behind and lay on her so that she was flattened against the rock and unable to move. Donald was in the mood for mating and Jo was taken aback. Nothing in all eighty-eight episodes of *Flipper* on the telly had prepared her for this. Fortunately, in her rubber suit, there was no chance of getting pregnant so, given no choice, she just lay back and thought, not of England, but her diminishing air supply. Aristotle dismissed copulation in dolphins as being 'neither very short nor very long'. But then he had never been a participant.

Local divers enjoyed many wonderful dolphin days. One spear fisherman was hunting for plaice buried in the sand. Donald soon caught on and went ahead like a beater, flushing them out with his nose. It was only when explosives were used during harbour alterations at nearby Port St Mary that he left the island for ever, to sojourn in Wales and Cornwall.

The Port Erin breakwater on which we dived wasn't cemented like a wall. It was assumed that concrete blocks one to two metres across didn't need to be assembled. Instead, they could be just dropped on the sea floor more or less in a line out from the shore. This quicker and 'cheaper' method cost £77,000, no small sum in the late 1800s.

The completed breakwater stretched 270 metres and terminated in a lighthouse perched on a massive concrete platform weighing 367 tonnes. To mark its completion a local benefactor, William Milner, of Milner's Patent Fire-resisting and Strong Holdfast Safe, held a party for the entire village, providing barrels of beer and roasting a whole bullock. The celebrations were premature. Just before Christmas in 1884 there was a gale. The grinding of the moving blocks could be heard over the rage of the storm, and the lighthouse and its platform toppled into the sea. The structure was wrecked and was now an effective breakwater only at low tide on calm days. But this jumble of submerged blocks presented an ideal collection of experimental plots, for it was easy to find a number of surfaces that were at the same depth and had the same aspect and slope or, if you preferred, that differed in any of these characteristics. It was the perfect field laboratory and conveniently situated right outside the marine station.

I was not the only one using the breakwater as a laboratory. Jo and Norman had observed that kelps dominated all the illuminated surfaces of the concrete blocks down to a depth of seven metres. But the lowest three metres of the breakwater were bare. Why?

They guessed that the edible sea urchins were to blame because they had seen them climbing up the kelp's stalks and biting chunks out of the fronds. Every month Jo and Norman removed all the urchins from the lower reaches of the breakwater. In the spring new kelp plants colonised the cleared zone, but not the urchin-infested region on either side. In the well-lit shallows the kelp had rapid powers of recovery, but in the dimmer depths they grew more slowly and the grazers had the upper hand.

Years earlier, in 1947, Norman had been the first to show that grazers were influential on the shore. He wondered whether young limpets moved into areas where there were few adults. So he removed all the limpets from a zone ten metres wide, stretching from the top of the shore down to low water mark. To his surprise, within months the entire strip was swamped with green seaweed. By removing the grazing pressure from browsing limpets he had saved millions of microscopic plants from being consumed and they grew to paint a green road on the shore. It was a vivid demonstration that herbivorous limpets ruled the roost.

On exposed shores limpets and barnacles had the upper hand, but on sheltered coasts the seaweeds took over. On in-between shores, the so-called balance of nature was a dynamic one, continually readjusting and reaching a new equilibrium in which seaweeds enjoyed a transient dominance before ceding to the limpets, only to regain the upper hand in later years.

After it forages across the surface of the submerged rock, a limpet always returns 'home' to the exact same spot to sit out the low tide. Each time it settles back down the rim of shell gouges out a circular ring from the rock and the shell becomes ever more deeply recessed into the rock. This gives it a lower profile so that it is less likely to be washed away by waves, and makes a better seal with the surface, helping to curb water loss when the tide is out.

Beneath the conical shell the animal is all 'foot'. It looks very like those suckers that used to tip the arrows for a small child's bow so they would stick on to a window or Grandpa's bald head. But it isn't suction that keeps the limpet in place. Instead, a thin layer of mucus gives a good seal and a tenacious grip. Those species that spend more time marauding around produce more slime, for it also lubricates locomotion. The rougher the terrain, the more slime they use to smooth their passage. But how does a one-footed animal walk on glue?

Slime has magical properties; under slight pressure it becomes slippery, but the instant the pressure is lifted it reverts to being sticky. Limpets and snails have waves of muscular contractions passing along the foot towards the head, pushing it forward. Each wave presses on the ground, liquefying the slime, which solidifies again immediately the wave has passed. Thus the animal glides forward on the slippery bit while the sticky parts prevent it from sliding backwards. This slip-stick progression allows it to climb even vertical walls and overhangs.

Producing a slippery carpet is exorbitantly expensive. Eighty per cent of the energy limpets and snails get from their food goes into making slime. This means that slithering costs ten times more than other animals expend running, swimming or even flying. As snails can still creep along even if their slime-producing glands are inactivated, why do they bother to smooth their way with mucus and devote thirty-five times more energy to the manufacture of slime than they do in contracting the foot muscles? Slime must provide other benefits.

When limpets go out foraging on the shore, they leave slime behind, like the silver shining trails on the garden path that trace the wanderings of last night's slugs. Just as Theseus unwound Ariadne's thread behind him in order to get out of the Minotaur's labyrinth, so the limpet unravels mucus to find its way home. But it doesn't always retrace the trail it made on the outward journey. The mucus trails of previous excursions persist for two months or more and any of them will do if they are going in the right direction. But how does the limpet recognise its own slime among the numerous tracks left by others, or which way leads to home? There must be directional clues in the slime that persist. In fact, so much information can be stored in organic films like these that they are expected to replace silicon chips in computers.

Limpets are slow-moving when feeding, but travel much faster when heading out to the feeding grounds. Often they slide out over existing trails. That way they don't have to produce as much new slime, so trail-following is cheaper than trailblazing.

These sticky ribbons of slime also trap settling larvae and algal spores and foster their rapid growth, especially if the slime has been spiked with nitrogenous fertiliser, as some limpets do. So it is worthwhile grazing on old trails as this not only recycles some of the expensive mucus, they are also seasoned with nutritious food plants. Some top shells even browse on whatever grows on their shell while at the same time smearing it with mucus to encourage further crops.

Slime makes the limpets' world go round. It helps them to repel enemies, and avoid drying out or freezing, and the blue-rayed limpet even blows a balloon of mucus to drift away on when moving home. There isn't a damp surface on Earth that doesn't have a film of slime clinging to it and offering a haven for all manner of microbes. At this moment trillions of tacky slimes all over the world are catching prey, cleaning lungs, attracting mates or easing the emission of waste.

Even the most fastidious human beings rely heavily on slime for many vital functions, but thankfully they retain it internally – at least most of the time.

Brace
Yourself

Port Erin, Isle of Man

Port Erin has a fine façade of seafront hotels that sweep up the hill following the curve of the sandy beach below. They look out over the bay and on the horizon the Mountains of Mourne in Ireland are plainly visible. It was June, but the beach was sparsely populated; a few hardy children screamed in the chilly shallows while their parents huddled behind windbreaks.

It had not always been so. In 1913 over 600,500 passengers descended on the island from June to August, twelve for every resident. They were mostly trippers from northern Britain. The mills of Lancashire closed down for their 'Wakes Week' and the shipyards were deserted during the 'Glasgow Fair'. The workers flocked to the Isle of Man. It was like going abroad. There was an exciting sea voyage on a steamer to an island that was not part of Britain and had its own money and postage stamps and a Manx parliament that dated back to the Vikings. It wasn't just a holiday, it was an adventure.

And there was plenty to do. Douglas offered pubs galore and a music hall where stars like Marie Lloyd performed, plus bawdy comics who lampooned the toffs. The flamboyant Florrie Ford once caught a banana thrown by a man in the stalls, swallowed it whole and invited him to come backstage later to get his skin back.

The middle-class visitors shunned the vulgarity of Douglas for the quiet pleasures of Port Erin. Arnold Bennett's Anna of the 'Five Towns' came here and fell in love. She enjoyed the informality and as the moon rose in the east over the little fishing village she thought it was the loveliest sight she had ever seen, 'a panorama of pure beauty transcending all imagined visions'. Excitement was provided by a proposal from an eligible man and a journey on the narrow-gauge steam train that 'annihilates' the fifteen miles to Douglas in a mere sixty-five minutes. For Anna the island was 'mysterious, enticing, enchanted . . . a remote entity fraught with strange secrets'.

Sensitive and romantic visitors were thrilled by the wildness of the Isle of Man: the endless tussocky moors on which stone circles were abandoned to the wind; the ground inhabited by the ancient dead; a Viking warrior with broken sword buried in his ship, together with a horse in full harness.

The charm still worked into the 1930s. Port Erin was described in a travel book as a place where 'health and sport reign supreme'. There were a dozen tennis courts on the south side of the bay and six on the north, in an area described somewhat extravagantly as 'the millionaire's playground'. The village also boasted the 'largest seawater baths in the British Isles' situated in a 'natural sun-flooded creek'.

The Marine Biological Station had a public aquarium and 'a single visit was calculated to awaken an unlooked-for thirst for knowledge'. It included a tour of the fish hatchery (the first of its kind in the British Isles), where plaice, lobsters and oysters were cultured from eggs. In a single year 50,000 home-grown lobsters and five million plaice were released into the sea.

On a weekend in August 1939 over 70,000 visitors returned home after the bank holiday. The local landladies little knew that there would be no more holidaymakers for six years. When war broke out the British government brought in landing permits making it difficult to sail to the Isle of Man. Trippers were out of the question. The desperate Boarding House Association put an advert in the London papers:

ISLE OF MAN
A really permanent place of security

And added, for the thrifty coward:

Safe accommodation to suit all pockets

It evoked little response and the islanders braced themselves for a lean war.

Meanwhile, in Britain, refugees were flooding in from all over Europe, mainly Jews fleeing from the Nazis. But they might have included spies and saboteurs. Something had to be done. In the spring of 1940 it was decided that all German aliens from sixteen to sixty would be interned for the duration of the war. As soon as Italy joined the conflict, all Italians became enemy aliens too. But where would these tens of thousands of people be held?

The Manx government wrote to the Home Office reminding them that the island had been the main internment camp during the First World War. Within months, boarding houses and hotels had been commandeered and seven internment camps established on the island.

There was certain to be some ill feeling towards Germans. Indeed, prior to that summer's TT motorcycle races, a local newspaper had proposed that riders on German and Italian machines should be banned. When the first refugees left Liverpool, the crowd spat at them and shouted 'Spies!' As they disembarked in Douglas, they were jeered and stoned by local youths. But landladies and hoteliers would receive a guinea a week for each lodger. These compulsory visitors would compensate for the lost holidaymakers.

The whole of Port Erin was encircled in barbed wire to house four thousand female internees, thus trebling the population of the village. The fence meant the locals were as much captives as the internees. One refugee from a concentration camp, on seeing the barbed wire, walked into the sea and drowned herself.

At first Jews and Nazi sympathisers were placed in the same hotels. Some of the women were vehement Nazis who celebrated Hitler's birthday and made large swastikas out of gorse. Those who took an oath of allegiance to the Third Reich received regular pocket money from Germany via the Red Cross. But most of the women were officially classified as 'friendly aliens'.

Many of the internees had relatives in concentration camps. It is ironic that those fleeing the shadow of the swastika should end up beneath the Manx flag that displays the 'Legs of Man', an emblem once thought to have been derived from the ancient swastika.

Some had even been born and brought up in Britain. One was an English pensioner who had married a German prisoner of war in 1918. He had never taken out British citizenship and therefore they were both classed as German nationals. 'Arve nevah set foot in Germany,' she protested in a broad cockney accent, 'and dan't know a bloody word of the language.'

Although permits were needed to enter and leave the camp, the women were free to shop, or swim and sunbathe on the beach. They were not allowed to talk to local men, but those who sunbathed in the nude must have got the men talking among themselves.

In Britain the papers griped about the liberal regime. The *Daily Telegraph* criticised the 'holiday atmosphere'. 'Aliens are in clover on the Isle of Man,' declared the *Belfast Telegraph*. It would get a little tougher

in winter, as the hotels were designed for summer visitors. Only the
ground floor had heating and coal was in short supply.

Many of the 'aliens' were educated professional women. They
included Kafka's last lover, Dora Diamant. Most felt insecure and
bewildered by the undecipherable Manx street names. Boredom was
rife. There was nothing to do. Some would knit a sweater then unpick
it and start all over again. Several became depressed, and were transferred to the local asylum; a few committed suicide. Many fretted for
their children who had been snatched away into care in Britain, and
the husbands they never saw even though they were interned elsewhere
on the island.

The 10,000 male internees were luckier. The energetic played
football in the inter-camp league, but there was also a generous
smattering of intellectuals, artists and performers. Rawicz and Landauer
gave piano recitals and the Amadeus Quartet was first formed here. Ten
artists from the camp exhibited at the Royal Academy's Summer
Exhibition. Kurt Schwitters converted his room into a studio peopled
by statues of quietly decaying porridge, dismembered chairs and stolen
lino squares. He never went visiting without taking a knife to liberate a
sample of floor covering to make linocuts. The jetsam discarded by the
sea and his fellow internees — seaweed and cigarette packets — was
transformed into collages of compelling beauty.

Some turned their hand to new skills. When a couple of male
internees came to the women's camp to replace roof tiles and fix the
plumbing, one was asked if this was his profession. 'No, I'm a gynae-
cologist,' he replied. 'But it's also a matter of tubes and pipes.'

An inmate dubbed his camp 'The best university in Europe'. The
camps also produced their own newspapers and the *Onchan Pioneer* was
a poignantly illustrated periodical.

In Port Erin, the skeleton staff at the marine station provided the
internees with an educational service and put on regular lectures, many

of them given by the internees themselves. English classes proved to be very popular. The library was opened to them and hundreds of new books on non-biological topics were acquired. Qualified internees carried out research projects on such varied topics as experimental wool dyeing, the teeth of a dinosaur and extracting fat from herring offal. Others went into business converting seaweed into garden fertiliser and waste food into chicken feed.

The camp commandant enlisted volunteers for laundry work, hairdressing, gardening and dressmaking. Many embroidered parachute silk and spun sheep's wool caught on the barbed wire. A shop in Port Erin now sold their knitwear and embroidery. The women's morale improved markedly when their children began to be brought across from Britain and they were reunited with their interned husbands in a married persons' camp in nearby Port St Mary. Among the new occupants of this camp were the distinguished German archaeologist Gerhard Bersu and his wife Maria. They were allowed to direct excavations of a Viking fort and an Iron Age farmstead, with armed guards keeping an eye on the volunteers. The acid soil had preserved the bases of so many posts that large quantities of ancient wood were used to cook sausages for lunch. With so much time and so few distractions, the excavations were comprehensive. 'Never again,' said Bersu, 'will a site of this type ever be excavated in such detail.'

Internees were being released and replaced by others all the time and substantial repatriations to Germany took place. Many were in tears and begged to stay. The Port Erin camp closed in September 1945, five years and four months after it had opened.

Many moved to the mainland and some, such as the publisher Lord Weidenfeld, the architecture historian Nikolas Pevsner and the business magnates Charles Forte and Tiny Rowland, went on to make their mark in Britain.

The hotels were inspected for damage and the owners compensated. One looked immaculate with all the furniture virtually unscratched, until they opened the drawers to find that only the drawer fronts remained, the rest had been carved into knick-knacks and sold to the locals.

After the war the tourists returned and in 1946 and 1947 visitors again topped 600,000 a year, but during the fifties their numbers progressively declined. Brigitte Bardot had taken off her clothes in *And God Created Woman* and sunbathing on the Riviera became glamorous. The south of the Isle of Man was called 'The Manx Riviera', but no one was fooled. The yucca-like leaves of 'Manx palms' gave a spurious air of exotic warmth, but they were really New Zealand cabbage trees and tough as sycamores.

Already one in seven went abroad and cheap package tours would soon tempt far more. In 1957, Benidorm was hardly bigger than Port Erin, and its unpaved roads were trodden by more donkeys than tourists. Two years later it had thirty-four hotels and four cinemas, and welcomed over 30,000 visitors. Within a decade it attracted more tourists than the whole of Italy.

The writing was on the sand for places such as Port Erin. Like all British seaside resorts, there's a short season in which warm days are possible, but the air is rarely hot and the sea always cold. Just when the day looks promising, a grey glove of fog fingers its way into the bay. The west coast of the British Isles gets reject weather from the Atlantic that blows in on the breeze. And what a breeze it is. Locals hang their wind chimes indoors for safety. A corner on the seafront has a handrail so you can haul yourself round. It is a place where hats fly and car doors jerk open so violently that many are pleated from losing altercations with the wind. I saw a woman catapulted across the road because she was holding on to a door caught by a gust. Even the tails have been blown from Manx cats.

The weather was always against British resorts. Sometime during the summer there was a blink of sun that we called the 'holiday season'. Their success in the past was down to marketing our inclement climate as a virtue. It was described as 'bracing'. This smacks of leaning into the wind, an attempt to extend the traditional stiff upper lip along the entire length of the body. Fresh air was good for you, whatever its velocity.

A more subtle invention was ozone. Ozone was first of all an odour. It was the 'smell of electricity' given off by early frictional machines that generated sparks. In 1840 Schönbein, the chemist who had invented gun cotton, was about to discover something that would prove equally popular. He isolated the gas and christened it 'ozone' (Greek for 'I smell'). How anything called 'I smell' caught the public's imagination is a mystery.

It had been assumed that the sea's emanations were unhealthy; the salty foam was the 'ocean's sweat', the tides, 'bouts of fever'. But like many new discoveries ozone became a fad. It is O_3 and heavier than air. What is effectively 'ozonised water' is well known to assist blondes as hydrogen peroxide, which was originally marketed as *Auricomous* (gold-making) *Hairwash*. But the 'ozone' that became all the rage in the 1860s and for decades after, was a mysterious property of seaside air. Nineteenth-century medics asserted that this 'heavy air', infused with salty vapour acquired as it blew over the wave crests, was 'vivifying' and essential to good health. They advised that as it was heavy it was vital to stay as low as possible to get maximum benefit. They failed to make it clear that this meant ankle height.

Ozone was a surprising choice as an invigorator as it is a powerful oxidising agent – don't look now, but your ankles are oxidising. It also attaches itself to organic compounds to make explosive ozonides. On top of all this it's a strong bleach and disinfectant, and corrodes cork and rubber in minutes. Trees like oaks and willows can generate ozone and damage pines growing downwind.

The aura of ozone persisted well into the twentieth century. In 1939 Blackpool boasted 'the healthiest ozone in Britain', following an earlier appeal to patriotism: 'Breathe British air at Blackpool'.

The bottom of Port Erin bay is carpeted with unattached seaweeds. Some have been torn from the rock, but most have just grown that way for they have no need of proper roots. Being unattached, they are flung ashore by storms to decay in heaps at the level of high tide. The terrible tang of rotting weed drifts across the beach on the breeze.

'Smell that,' said a father to his offspring. 'That's ozone,' he said knowingly. 'Take a deep breath, does you the world of good.'

Perhaps he wasn't aware that the easiest way to get a fix of ozone on a sunny day was to inhale close to a car exhaust.

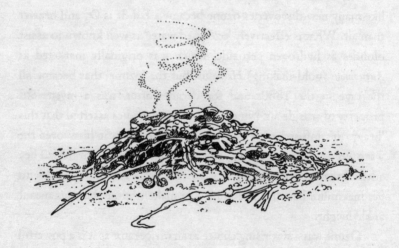

Green

Glasgow, Scotland

After finishing my PhD, I had an interview at Glasgow University. The train's lavatory bore the usual warning, *Do not flush when the train is standing in the station.* Underneath some wit had scribbled *Except at Glasgow Central.* This reflected the public's view of the city.

I stayed overnight at the Central Hotel where, years before, a plague-carrying rat had been discovered. This so intrigued the physiologist John Scott Haldane that when in Glasgow he never stayed anywhere else.

The name Glasgow probably means 'green hollow' from the Gaelic *glas cau*, pronounced 'glas kay', the way many Glaswegians still pronounce Glasgow.

By boat Glasgow is one hundred and ninety kilometres from both the Irish Sea and the Atlantic Ocean, yet it is a city of the sea. In the eighteenth century it was most famous for its salted herring known as 'Glasgow magistrates'. Then Glasgow began trading with the American colonies and in 1772 alone hundreds of ships brought in 20,000 tons of tobacco plus 180,000 gallons of rum and huge quantities of sugar from the West Indies; there is still a Jamaica Street downtown. Tobacco barons built churches to thank God for rewarding their humility.

St Mungo, the city's patron saint, had declared, 'Lord, let Glasgow flourish through the preaching of the Word and praising Thy name.' 'Let Glasgow flourish' became the city's motto, suitably pruned to give more scope to the means. As the economic dynamo of Scotland, Glasgow became the centre of the textile industry. Armies of workers spun, wove, dyed and printed imported cotton. Bleaching and dyeing of cloth required chemicals and Tennant's St Rollox Chemical Works became the biggest in Europe. Its chimney generously disgorged a rich menu of noxious fumes over a wide area. By the 1930s, fifty tonnes of soot fell on every square mile of the city each year and smoke cut out so much sunlight that the locals lived all day in crepuscular gloom.

The 'dear green place' had been sullied and the Gorbals, once a leafy suburb, became a byword for slums. When I arrived in 1966 the city was just beginning its rehabilitation in the public's esteem.

I was appointed Assistant (Junior Lecturer) in the Department of Botany. Glasgow University was founded in 1451. That was twenty-four years before the first book was printed in English and forty-one before Columbus discovered the West Indies. More importantly, as any Glasgow graduate will tell you, it was 131 years prior to the founding of Edinburgh University.

Samuel Johnson said, 'Much can be made of a Scotsman, if he be caught young,' but feared that those 'bred in the universities of Scotland obtained a mediocrity of knowledge between learning and ignorance'. I had been hired to contribute to one or the other.

The Botany Department had spent several centuries accumulating stuff in cupboards. Planned renovations meant it was time for a clear-out. Among the beautiful brass microscopes, correspondence between famous Victorian naturalists and the notification of Darwin's funeral, I discovered four lovely watercolours of flowers. I gave two to the prof's secretary for her office and kept two for my room. It was some years later that Charles Rennie Macintosh was resurrected from obscurity and I realised that I had his paintings hanging on my wall. I notified the university and within days they had been reclaimed. Macintosh had lived in a house nearby, and after his death the university had acquired it complete with his decor and furnishings.

I was affiliated with Elsie Conway who also worked on marine algae. She was a 'fine block of a woman', although many saw her as a battleship that steamed forth to engage the enemy. Her main armament was a booming voice, so that on the shore even in a force nine gale she could still be heard by the students. I witnessed her explaining the stresses of living in the splash zone when, without warning, a wave drenched her. Her lecture never missed a beat and her soggy state perfectly illustrated the point she was making.

She had a flair for the accidentally outlandish. When a class was searching vainly under the microscope for signs of reproduction on an alga, one lad found something. 'Class!' she bellowed. 'Come and see this boy's sex organs.' For weeks afterwards, whenever girls passed him, one was certain to ask her friend, 'Have *you* seen his sex organs?'

Her son referred to Elsie as 'Her Botanical Majesty'. Although warm and friendly, she brooked no nonsense. For many years she was the warden – I almost put warder – of a women's hall of residence. Then it became mixed. The boys were put in the newly built Conway Tower, but when she realised they could look down into the girls' rooms she trouped them out and moved the girls in. So the females acquired a surfeit of shaver sockets and the lads an abundance of wall-fixed hygiene bins. She was unconcerned that the boys might romp naked at their windows – because none of her 'gals' would be looking.

The first time I assisted her in a practical class she warned me in advance not to correct her. I never dared to do so. She was one of several female botanists who had been drafted into marine biology during the war. When the Japanese entered the fray our supplies of agar, a jelly derived from seaweeds, were cut off. This was more serious than it sounds because agar was an ideal medium for growing the fungi that produced the newly developed antibiotics. As most of the men had been conscripted, lady botanists were sent out to find adequate supplies of red algae that could produce home-made agar, and that's exactly what they did.

I was pretty green, never having taught before, but was thrown in at the deep end. I began with the class of first-year biologists – five hundred of them – and shortly afterwards was given an additional three hundred pre-clinical medics and dentists. The dentals were no trouble – they slept most of the time – but the medics were chafing to get on with cutting up corpses and patronising patients on the wards. To have to teach them about ecology and the environment was dangerous. The cream of Glasgow's young lecturers, brimming with confidence, had gone over the top to face this enemy with only a stick of chalk. Some came back shell-shocked, broken men, others were never seen again.

The students' main weapons were jeering and stamping their feet in disapproval. To be under a continual barrage of hundreds of stamping feet day after day must have been a terrible experience. One of the big lecture theatres had enormous reversing mirrors on either flank so that the lecturer might feel able to turn his back on the audience to write on the board without losing sight of the foe.

Another problem was paper aeroplanes. If the students became bored, the lone lecturer was inundated by wave after wave of aircraft. At the end of each lecture two porters with special brooms a metre across went in to sweep up the heaps of crashed planes. For my first lecture I went armed with a starting pistol in my pocket. To add to the tension, President Johnson had ordered air strikes against the North Vietnamese. For twenty minutes there was nothing, then out of the blue came a solitary paper plane. I whipped out my pistol and shot it. The noise was deafening and then there was the silence of surprise. In a loud voice I declared, as Ho Chi Minh had done, 'There will be no further talks until you cease the bombing of the North.' The entire class cheered and applauded. I wondered if they might launch more planes just to see me shoot at them, but it never happened.

For recreation I was persuaded to join the university badminton club. Being handy at tennis, I thought badminton would be a cinch. Unfortunately, when I delivered my ferocious overhead smash, the shuttlecock plummeted to Earth, its feathers in disarray like a broken budgie. By the time I learned to flick the shuttlecock it was too late; I had already lost to a man with a false leg, although it has to be said that his remaining leg was first class. He ran little, but was an accomplished leaner.

Such a defeat might have driven me to drink, but Glaswegians took their drinking too seriously for me to compete. In a bar a gloomy man turned to me and said, 'I'll be glad when I've had enough.' There were even those that snitched a bottle of milk from a doorstep and then bubbled it with fumes from one of the old gas brackets that had illuminated the landings on tenement stairs. The resulting brew was called 'stairheed champagne' until French winemakers asserted that to be called *champagne* it had to be gassed from a bracket in Reims.

Win made regular trips up to Glasgow and we shared a tiny single bed in my one-room flat. One morning the landlord came to inspect the premises. I managed to show him around the room like a matador playing a bull and he left without noticing there was a naked woman in the bed. Win got fed up with this intermittent romance and made it clear that I was perilously close to a no-Win situation. Within a year we were married.

The flat was in Cecil Street close to the campus. Its tenements were cliffs of once pale sandstone, but decades of soot and rain had dulled their sparkle. The front-door steps sprouted vegetation and the cracked downspouts had disagreements about where the rain should go. Yet the more the street decayed, the more it came to life. Every week coal lorries ambled past with a man walking alongside shouting his wares. So many housewives offered their favours for a bag of nutty slack that the weary coalman shouted 'Coal for money!' to make the terms clear.

If you glanced out of the window any night of the week something was going on. Chairs and tables out on the street are called a café in Paris and an eviction in Glasgow. Even more common was to see a couple toting twenty carrier bags piling into a taxi and driving away from their flat. This indicated they had failed to pay the rent, and was called a moonlight flit.

Many cities have germinated on the banks of a river. Over 30 per cent of the people in Britain live within thirty kilometres of an estuary and no one lives more than a hundred kilometres away. Glasgow is situated at the highest inland reach of the tide. Thus, from the beginning it had fresh water on the one hand and access to the sea, and therefore trade, on the other.

Glasgow was sited on a ford in the river. When Daniel Defoe crossed, the water was 'scarce over the horse's hooves'. But a summer trickle often became a winter torrent that flooded the city.

When the tobacco trade flourished, the transatlantic ships had to dock at Port Glasgow, twenty-four kilometres downstream. The bulk cargoes were then broken down into manageable lots to be carted or shipped up to Glasgow in small boats on a rising tide. It was patently inefficient, so engineers were called in. Initially, they narrowed the channel to make it deeper, but still the city flooded. Later, with more powerful dredgers, they were able to both deepen and widen the river so that more sea water came to scour the bed and flush out the sediment. This also allowed excessive flows of fresh water to pass through without submerging the town.

Not only could ships now reach to the heart of the city, they could be built there. James Watt devised the steam engine at Glasgow University; *Comet*, the first seagoing steamship, was constructed here in 1812, and the first iron ship in 1841. Soon 90 per cent of all the world's ships were built in Glasgow. A brass plate saying 'Clyde-built' became a symbol of craftsmanship. In addition to ten large shipyards, there were several structural engineering companies, making girders, building cranes and bridges, and four railway works supplying locomotives to the British Empire. The famous ships born here included the Tsar's magnificent steam yacht, *Livadia*, so stable you could play billiards at sea, and most of the great Cunard liners: *Lusitania, Queen Mary, Queen Elizabeth*.

And as the population grew so the river deteriorated. It was expected to carry away their sewage and rubbish, but a price was paid. By the 1850s, the river's odour drove passengers from ships to the railway. Every major river in Britain was just as bad. In the long hot summer of 1858, the nasally tortured MPs in the Houses of Parliament debated the 'big stink' from the Thames and voted several million pounds to solve the problem. If you fell into any of our great rivers it was considered kinder to let you drown than to haul you out to await cholera or typhoid.

At the end of the eighteenth century the Clyde had sustained several commercial fisheries, and the salmon and sea trout runs extended beyond Glasgow, but within sixty years the entire estuary was devoid of fish.

In 1894, sewage treatment began and in sixteen years three major sewage works were constructed. The solids that settled out were shipped daily to a dumping ground off the Isle of Bute in the Firth of Clyde. It was easy to find it by simply counting the number of tomato pips in samples of bottom mud; once they exceed 10,000 per square metre, you were there.

When I took my first voyage down the Clyde in the summer of 1967, I was driven from the open deck by the smell of bad eggs. There was also so much methane being given off that the river was a fire risk.

Thanks to sewage treatment it was soon on the mend. Fish began to return a species at a time. By 1985, nineteen species reached as far as the city and salmon were breeding again in the river. Recently, a minke whale explored the estuary and dolphins were seen in the heart of Glasgow.

I like rivers. They seem to know where they're going. My eyes always follow the flow downstream towards the sea. As the river widens it seems to have second thoughts about its impending encounter with the ocean; the flow slackens and the water drops the burden of silt it has been carrying. Even as late as the early 1970s the resulting mudflats were highly enriched with organic matter from sewage. This fed enormous populations of mud-dwelling crustaceans and worms, which in turn sustained tens of thousands of lapwings and oystercatchers plus the largest single gathering of redshanks in Europe, which consumed 40,000 crustaceans a day. Someone always benefits, even from pollution.

In the 1980s, the numbers of lapwings and redshanks fell rapidly. As the estuary was cleaned, the muds received less organic matter so the armies of invertebrates declined, depriving the birds of food. In addition, juvenile flounders that had recolonised the estuary now took their share of the mud dwellers, leaving less for the birds. On the other hand, cormorants, which feed on flounder, increased and one was recently seen close to Glasgow city centre.

The human population of Clydesiders ebbed and flowed with the fortunes of the river. There were repeated slumps of the shipbuilding industry and in the depression of the 1930s most yards did not have a single order on their books. Even the half-built hull of the mighty *Queen Mary* lay rusting in the dock for years until a government subsidy allowed it to be completed.

Within forty years only one prestigious contract was left – the construction of the *QE2*. The great shipbuilding rivers, the Clyde and the Tyne, were simply not wide enough to launch the giant tankers and oilrigs that would be the main business from now on. The industry had blossomed and would wither. By the end of the 1970s all the major yards were gone.

When I first saw the Clyde, she was already becoming a place of empty slipways and idle cranes. The only boats that made it up to Glasgow were police launches and the occasional dredger. But amid the abandonment, Buddleia, ragwort and rosebay willowherb were taking over. The regreening of the banks of Clyde had begun.

In the Zone

Millport, Scotland

At its mouth the river opens out into a huge drowned valley. The Firth of Clyde is ninety kilometres by fifty wide and rapidly deepens to over ninety metres. Tentacular sea lochs probe inland until far from the scent of the sea. There is a scatter of islands of rich red sandstone baked in a Permian desert: Arran, Bute and Cumbrae. Until the outbreak of the First World War, forty steamers a day took Glaswegians 'doon the watter' to Millport on Cumbrae, or Rothesay on Bute, christened the 'Madeira of Scotland' by an enterprising local who had drunk too much Madeira wine without visiting the vineyards. Although Cumbrae had been a resort since the eighteenth century, it had few claims to fame. Even a guidebook could only extol the excitement of visiting the grave of the nineteenth century minister James Adam, who 'used to pray for the Great and Little Cumbraes and the adjacent islands of Britain and Ireland'.

It also had the Millport Marine Station, which had originally been a 'houseboat' moored to the shore. Its name, the *Ark*, reflected its antiquity. It was open to visitors and a sign stated: *In order to make the education value of the Institution as largely available as possible, a nominal charge is made for admission, viz., ONE PENNY.*

Amazingly, this floating laboratory also attracted a stream of distinguished researchers, until it broke up in a storm in 1900. By then it had been replaced by a permanent building. It was on the adjacent shores that my students did their research. We concentrated on the region between the tides; the fluctuating boundary between land and sea.

Maoris believed the rise and fall of the water came from the breathing of a giant who slept on the bottom of the ocean. I once heard a child ask, 'Mummy, who's pushing the water?' I wanted to tell her it was the moon and it was pulling not pushing. The gravitational tug of the moon is so great that an ocean liner weighs seven kilograms less under a full moon. No wonder our attractive moon causes the mobile ocean to form an enormous hill of water at the place immediately beneath its spell and, because of centrifugal force, simultaneously on the opposite side of the world.

As the earth spins, the water follows the moon and a tidal wave rushes across the planet at 1,670 kilometres per hour. When the wave reaches our shore we prosaically call it high tide. No consecutive tides rise to the same level, but there is a pattern.

Although the sun is four hundred times further away than the moon it is also four hundred times bigger, which is why they appear to be the same size in the sky and overlap exactly at an eclipse. It means that the sun's gravity also affects the tides, sometimes augmenting the pull of the moon (so-called spring tides), at other times counteracting it (neap tides). At Millport the vertical range of the spring tides is half as big again as that of neaps. Spring tides recur every fourteen days, when the moon is full or new.

This regular cycle of larger and smaller tides creates different regions on the shore; the lower reaches are only exposed to air during springs, the mid shore is wetted and dried with every tide, the upper shore is only reliably wetted during springs, and the area above this is entirely reliant on splash from waves. These zones attract different inhabitants.

A zoologist called Alan Stephenson toured the globe examining the zonation of shore dwellers. In Australia and Bermuda, California and South Africa, he saw similar patterns and similar types of animals occupying different regions of the shore. The mid shore was dominated by barnacles and higher up, in the spray zone, winkles and lichens thrived. Seen from a distance, exposed shores display a horizontal white band of barnacles topped by a broad stripe of black or orange lichens. There are stories, perhaps apocryphal, of Stephenson getting so good at surveying shores he didn't need to set foot on them. Occasionally he was said to scan the shore with binoculars from the cliffs above and dictate to his wife: 'Barnacles common, periwinkles sparse, top shells . . . seemingly absent.'

He was also an authority on sea anemones and a talented artist. His classic book on the British sea anemones is adorned by beautiful paintings of them and delightful vignettes of a naked sea nymph prancing in the surf and reclining on rocks. The nymph is clearly his wife Anne. Librarians at his university had inked out these vignettes in case she induced a prurient interest in anemones.

On sheltered shores, seaweeds obscure Stephenson's 'universal' patterns and create zones of their own, each dominated by a different species. These plants occupy a halfway world but, like almost all shore dwellers, owe their allegiance to the sea. They are truly marine and grow best when submerged, yet some live high on the shore, where submergence is brief and occasional. When abandoned by the tide, barnacles and mussels clamp their shells to conserve water, but leave a tiny gap so they can breath. The intertidal wracks have no such protective cover. My American research student, Mark, showed that, unlike cacti growing in arid places, they have neither internal reserves nor any means of conserving water. On the upper shore, where the tide might not return for a week, the plants desiccate and their internal processes virtually close down for the duration.

Survival on the shore requires an acceptance of stress. Fortunately, most seaweeds growing high on the shore can tolerate being dried out for far longer than those characteristic of lower down, and are more adept at getting back into action when resubmerged. Channelled wrack, which lives at the very edge of the spray zone, becomes so dry that if crushed in your hand it crumbles to dust like a roughly handled mummy, but if instead it is placed in sea water, it rapidly swells back into its old rubbery self and resumes photosynthesis.

The shore is probably the only living community that is usually examined by ecologists only when all its inhabitants are out of their element and out of action. When the tide retreats most shore dwellers hunker down and wait for its return. To go snorkelling over the shore at high tide is to discover a different world. The flopped blanket of weed is held aloft as a field of lazily swaying plants. The previously static limpets slide across the rock tirelessly scraping up tiny algae, the barnacles now rhythmically flick out their feeding legs like a fluttering eyelash. Crabs come out of hiding, winkles are awake and climbing up the weed.

Being submerged is also not without its dangers. When the tide comes in, predators emerge hopeful of homicide. The dog whelk has a file in its mouth to drill holes through barnacle shells. This can be a long business. It rasps away for five minutes or more and then squirts acid into the dimple, allows half an hour for it to soften the shell, and has another go with its file. Days may pass before it eventually cuts a keyhole through which death comes calling. It then dissolves its victim alive and wrenches off morsels until the barnacle dies of attrition. It would be simpler to have a go at smaller, easier-to-drill barnacles, but in terms of energy gained for energy expended the whelk should go for the big ones, which is exactly what it does. The lives of most animals are constrained by the need to glean the most food for the least effort and at minimum risk of being eaten themselves. The humble snail is as adept at cost-benefit analysis as any economist.

The other snails that commonly roam on rocky shores are periwinkles. Unlike the whelk they can't drill. Instead, they have a rasp with hundreds of rows of teeth to scrape algae from the rock or excavate the fronds of seaweeds. David, another of my students, showed that they are fussy feeders. If offered a choice, they invariably chose one type of seaweed in preference to the other. The common winkle loves sea lettuce and when plants are washed up on the shore it jumps aboard like a sailor on a single girl. If none are available, it will snack on most of the brown wracks, but would rather starve than eat knotted wrack, which often covers the mid shore.

In a maze the snails move towards a source of sea water that has been in contact with sea lettuce, but move away from any similar traces of knotted wrack. If the snails are idling in an aquarium tank, a single drop of 'sea lettuce water' sets off their feeding reflex and they rasp away at non-existent plants. When they are eating genuine sea lettuce, they can be halted in mid rasp by a drop of 'essence' of knotted wrack.

Plants are at the mercy of the animals that consume them; grass can't hide from the cow. Sea lettuce relies on growing rapidly, reproducing early on and then dying away only to spring up anew elsewhere. In this way winkles may locate a patch of appetising weed, but it is insufficient to sustain them long-term. So the number of grazers doesn't build up in one place to the detriment of the plants. But this 'here today and gone tomorrow' strategy would not work for knotted wrack as it lives for many years. It needs to defend itself. Like many land plants assailed by voracious insects, wracks have an abundance of distasteful tannins, and knotted wrack, which often dominates the mid-shore, has an abundance of the more astringent ones. To economise on the production costs of protective chemicals, it only raises its defences if attacked.

When I look at the vignette of Mrs Stephenson lolling on the shore, I confess that I think more of the chemistry bubbling inside the surrounding plants than that generated by the naked nymph.

Go West Young Man

Argyll, Scotland

At last I was able to begin my first independent research. Scientists thrive in teams and no one doubts that collaboration leads to synergy. But there is a special thrill in converting a novel idea into a research proposal and persuading a research council to fund it – and doing it all yourself.

Another research grant was used to buy diving equipment. The local dive shop was very reasonable and jokingly offered free ascents. John Milburn, a physiologist whose hobby was diving, became my main companion on a series of expeditions to the west coast of Scotland.

In Argyll, the scowling peaks are lined with age. They were once almost as high as the Andes, but have been worn down by time. The now deserted hills bear evidence of past inhabitants: standing stones, underground passage graves and huge funeral cairns. On slopes overlooking the sea are neolithic graffiti: enigmatic cup and ring marks, spirals and mazes pecked into the shining wet schist. Peat diggers have found ancient bodies in the bogs, so perfectly preserved that the police took their fingerprints to see if they had a criminal record. Time moves slowly here. A stranger who came to the Isle of Coll asked a local what time it was. 'August' was the reply.

In winter, the two coasts of Scotland are like beers: the east is bitter, the west is mild. In the summer, curtains of rain unfurl before you then drop on your head. When we couldn't possibly get any wetter, we went diving.

Thirsty moors swallow the deluge to make soggy valleys and great raised mires, sponges of bog moss that dome above the water beneath. Some bounce like a waterbed as you walk across them, not knowing how deep is the lake beneath.

Although John and I planned and organised the trips, Elsie Conway occasionally came along and got into the habit of saying, 'Let's put the boys down here.' We often dived for several hours a day and drove between sites in our damp neoprene suits. When we finally changed, we checked for sprouting barnacles and feared athlete's foot up to the armpits.

For several years from June to November we roamed the west coast dipping into dozens of sites like avid readers who had found the lost library of Alexandria. From every shelf we sampled the voluminous fauna and flora and filled net shopping bags with wonderful finds.

At Dunstaffnage a once mighty castle decays. It was home to Jacob's pillow, the magical Stone of Destiny, which had been stolen from Ireland and was then filched by the English to slot under the coronation chair in Westminster Abbey. Dunstaffnage means the 'Fort of Seaweed Point' so we plunged into the tidal torrent in the gully separating it from Sheep Island, although only when the sea was resting between tides. We also dived off wave-pummelled headlands such as the aptly named Dun Mor (Great Fortress), where at the lowest margin of the shore streamlined seaweeds called *Alaria* ('winged') flew back and forth in the surf. We fought the waves in our little inflatable propelled by a puny Seagull outboard motor less powerful than an electric egg whisk. Even when diving fifteen metres down we were violently flung back and forth a couple of metres. We discovered that kelps avoided the chaos of the shallows by growing much deeper than they did at sheltered sites and, because the Atlantic water is so clear, the kelp forest could penetrate down to twenty-four metres.

Then there were the quiet shores of the long sea lochs where the peaty water eased towards the sea. Only nine metres down in Loch Feochan the visibility was less than a metre and even the most shade-loving weeds could live no deeper. In its feeder rivers salmon were leaping during the August run. While I was chatting to a local fisherman he caught a magnificent fish. 'That's a great pity for I no have a licence,' he admitted sorrowfully. 'I'll have to throw it back.' So he threw it in the back of his boat.

The water in some of the lochs was five or six degrees warmer that the sea, so barnacles grew faster and ripened earlier than on the open coast. Under every rock a gang of sandhoppers congregated and when exposed, scattered like schoolboys caught smoking behind the bike shed.

Seil is an island, but only just. Clachan Sound, which separates it from the mainland, is never more than twenty metres wide and mostly much less. Within its wooded banks a river flows, but this stream is the tide. A bridge designed by Thomas Telford resembles one of the Loch Ness Monster's humps and spans the sound with five metres clearance. Postcards call it the 'Bridge across the Atlantic'.

Occasionally, whales swim into the shallows of the sound and cannot reverse. In 1835, a twenty-four-metre animal died here and two years later 192 pilot whales met a similar fate.

The banks are festooned with wracks, all neatly zoned, but confined to a vertical band of only three metres, for that is the range of the tide. On the open coast at Dun Mor the span occupied by shore dwellers is three times greater, a measure of the extent to which waves and spray can extend the influence of the sea.

Off the seaward end of Seil is Easdale, a tiny islet pocked with desolate slate quarries. They are reminders of a time when this area had a great industry. In the 1770s, two and a half million Easdale slates a year were dispatched to England, Norway, Canada and the West Indies. They roofed Glasgow and Edinburgh and even New York, for slates were profitable ballast for transatlantic freighters.

For three hundred years quarrying brought intermittent prosperity and employment. But they cut away so much rock that in 1881 the thin side walls were breached by a storm and the Atlantic poured in. Two hundred and forty men lost their jobs and that was just the beginning. The high cost of shipping and competition from new lighter roof tiles made the quarries uneconomic. When we came the last one was about to close.

The flooded quarry was a wonder. Imagine a tide pool as big as a swimming pool, but sixty metres deep. It was like diving into an immense chimney, a vertical cavern leading into the dark heart of the island. The water was transparent, but down below the stillness was unnerving. There were more species here than anywhere else we dived. The walls were vertical with terraces at intervals. On screes of slate fragments, some fanned like playing cards, green seaweeds in the form of human hands protruded, a weird sight in the gloom

Deeper down, silt had accumulated, and by the autumn the lack of flushing waves and currents meant that all the oxygen was used up in the water below twenty-four metres and almost everything died. When we came up, our mud-smeared diving suits stank of bad eggs — the smell of a world without oxygen.

This ancient rock has few fossils, but is not too old to bear memories. Everywhere there were rusting fixtures where machinery had once been, and skilled workers had removed the slate and strong men had pushed it up into the light in wheelbarrows.

*

The Hebrides comprise over five hundred islands not separated but joined by water, as the locals say. Some, like Seil, are so close you can swim to them, and others, austere and wind-whipped, lie almost two hundred kilometres out into the Atlantic. In the shadows of their hills and mountains are 7,580 lochs, some so remote that their fish have never sniffed an angler's bait.

Many of the islands were forged in fire, and volcanoes last active 60 million years ago now masquerade as the islands of Mull, Rum and Skye. Lava covered the land over a kilometre deep in places. Here and there on the shore are circles on the rock where bubbles in the lava burst like boiling porridge. Perhaps there is an echo of this in the legend of Iona, where fire-breathing dragons emerged from the sea leaving trails of opalescent slime.

There is a famine of wood. Most of the islands are treeless and peat-darkened. When the magnificent stone circle of Callanish on Lewis was discovered, it had to be dug out from five metres of peat. The outer islands were thought to be remote and dangerous, at the very edge of the world. They were fortress outposts for Vikings, then rival clans – a place to be avoided. This all changed after the defeat of the Jacobites at Culloden in 1745, the very last battle to be fought on British soil.

After the battle, Bonnie Prince Charlie fled not just over the sea to Skye, but from one lonely island to another by rowing boat at night. He spent five months sleeping in caves and crofts and hiding in the heather until he was rescued by a French ship and escaped into legend. The British government was determined to capture him and offered a massive reward of £30,000. It was the largest manhunt in history. Troops were diverted to hunt him down in the glens and the navy dispatched to search the Hebrides. They were in uncharted waters and had no guidance through the dangerous maze of skerries, whirlpools and barely submerged fangs of rock.

The British were determined that this fiasco would not happen again. If they had to pursue the unruly Scots they must have maps to help them. Thomas Telford was one of those sent to survey the mainland and the army's engineers built roads fit for moving troops rapidly to quell any future discontent.

The task of charting the west coast fell to Murdock Mackenzie. He was an unusual fellow – a mathematician yet a man of action. He was also an experienced cartographer who was already engaged on a geographical and hydrographic survey of Orkney. He and his assistants cruised the Hebrides navigating without a chronometer and determining longitude by observing the occasional eclipses of Jupiter's moons.

He established a three-mile-long baseline in the Outer Hebrides and from its ends measured the angles to selected high points on all the islands he could see. He laboriously climbed every one of these peaks, built cairns and from these vantage points triangulated others and meticulously sketched the coastline he could see below. Soon his cobweb of triangles covered the islands.

It was a long job; he and his helpers spent from March to November every year sailing the Hebridean seas and climbing the precipitous hills. The task carried great responsibility, for lives depended on the accuracy of sea charts. In the winter he busied himself plotting all the readings and marrying them with his sketches. He began in 1748 and ten years later he had charted the entire west coast of Scotland, both islands and mainland; a coastline thousands of kilometres long and double that of France. After he retired he took a further decade to produce a magnificent set of one-inch-to-the-mile charts of the entire west coasts of Britain and Ireland.

Had Mackenzie lived two hundred years later, he would not have been content with his soundings of the water's depth with a shot line, but would have been intrigued by our underwater exploits. I suspect he would have plunged into the sea to devise methods for surveying under water, and would have made a far better job of it than we did.

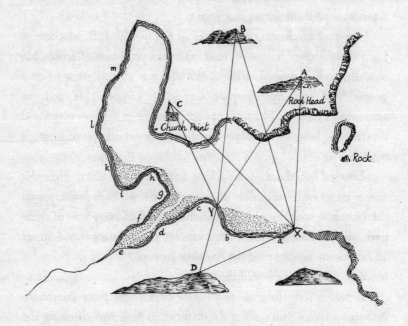

What's in a Name?

Hebrides, Scotland

The names of the islands are as bold and rugged as the places themselves: Coll, Canna, Raasay, Muck, Jura. Visitors described the terrain they saw. First came the Celts, then the Norsemen for whom the Hebrides were the Sudreyjar or Soudreys, the southern islands. Even today the head of the Protestant Church on the Isle of Man is the Bishop of Sodor (= Soudreys) and Man.

And so they labelled the land: Coll, 'Hilltop'; Canna, 'Pot-shaped'; Eigg, 'Cleft', perhaps because of its central ravine; Mull, 'Snout', after its promontories; Muck, 'Isle of Swine', and Staffa, 'Pillar Island', for that cathedral of basalt columns. Because the names have been whispered down the ages, the origin of some has been obscured. Skye supposedly means 'wing' but nobody now knows why. Lewis is old Norse for either 'Silent House' or perhaps 'House of Song'. Today it depends on which day you visit; 'frivolous' music is frowned upon throughout the week, but hymns are encouraged on Sunday.

Most place names originally had a meaning even if today it has been lost. We are happy to accept names as being abstract labels. How many tourists to Las Vegas think of it as 'The Plains', or Los Angeles as 'The Angels', or consider how today's 'twenty suburbs in search of a city' could have evolved from the original village of El Pueblo de Nuestra Señora la Reina de Los Angeles (The Town of Our Lady the Queen of the Angels)? Does the English publican know that the Elephant and Castle is a corruption of *infanta de Castile* or that the Cat and Fiddle was originally *Canton de Fidele* in memory of an English knight who held Castile against the French? Do the residents of streets such as Elm Drive or Cherry Tree Gardens ever wonder where the trees are or whether there were any trees there in the first place? We simply take names for granted.

Even our own names are thought of as useful but meaningless appendages to the people we know. When christening, do we consider the suitability of the label we inflict on our innocent babes? Perhaps Fergus is indeed a 'manly choice' and Donald might become a 'world leader' and Saddam 'one who confronts', but will shy Bronwen be pleased to be known for her 'white bosom'? Do we really want Deirdre to turn out to be the 'raging one' or Camilla to be 'attendant at a sacrifice'?

We may be nonplussed by Polish names, but manage to cope easily with unfamiliar or involved ones in English such as Bartholomew, Ermintrude, Jacqueline or Maximilian. Why then do we panic at the no more forbidding proper names given to living creatures? After all, Latin and Greek suffuse our everyday language.

A significant advance in the organisation of biology was the proposal by Linnaeus to give every species a double-barrelled name. It should be in Latin or Greek of course because they were the languages of learned men across Europe, although other languages were allowed to infiltrate; *Jasminum* is Persian, *Alchemilla* Arabic and *Anemone*, *Crocus* and *Rosa* are Oriental.

The idea was to provide universally accepted names and avoid confusion. It would also help to order the world by formalising affinities so that we would appreciate the relationship between one organism and the next. The first name would give the genus and be common to all close relations, the second name would be particular to that species. So the dog is *Canis familiaris*, the dingo was *Canis dingo* (dingo was the Aborigine name for it). It was Linnaeus himself who first designated the wracks of the shore: the saw-edged seaweed is *Fucus serratus*; the spirally twisted one, *Fucus spiralis*; and the one with bladders (vesicles), *Fucus vesiculosus*.

Inevitably, biologists became obsessed with classifying the living world, for to name something was to acquire some small command of it. Much of the effort in biology was devoted to the search for new species not just to reveal the diversity of God's world, but to gain the prestige of having conferred a name. This achievement is displayed alongside the creature's name. So *Natica alderi* (Forbes) indicates that Edward Forbes was the name-giver for that species of snail. In Victorian times not to have described a new species was a stain on your career. An accomplished biologist was dismissed after his death as 'never having named a new species of animal [or] proposed a new generic term'.

There was therefore a strong temptation to split perfectly valid existing species into two new entities and then name them. In 1838, a taxonomist complained of those 'who take *pride* in the manufacturing of names . . . hoping that by thrusting a Jack-Scroggins-of-a-name into notice, it will be handed down to posterity'. Another critic, exasperated by those who 'hurry to be first to the species market', dismissed them as 'Covent gardeners with their . . . unripe strawberries, a shilling a dozen'. Charles Darwin was 'rabid against species mongers' and complained that some taxonomists acted as if they had 'actually made the species, and it was their own property'.

Fortunately, etiquette forbade them from naming a species after themselves, but it did no harm to your career to commemorate your patron or a fellow taxonomist who might return the compliment with his next discovery. Often, of course, this was a genuine tribute to a mentor. *Ophiocomina goodsiri*, a brittlestar, was Edward Forbes's tribute to the professor of anatomy John Goodsir, who was the first person to befriend him on his arrival at Edinburgh University (and would also be the last to leave his deathbed twenty-five years later).

Many of those honoured had no biological credentials. Louis Antoine de Bougainville was a soldier and distinguished mathematician, but it was for his circumnavigation of the world that he was commemorated for ever as beautiful *Bougainvillea*. Perhaps Sir Thomas Stamford Raffles, the founder of Singapore and London Zoo, was less impressed by being immortalised as *Rafflesia*, the corpse lily that stinks of carrion. Scorn was heaped on those who erected names commemorating 'persons of eminence . . . Peruvian princesses or Hottentots'. Approval would probably have been withheld for the name of a cave-dwelling beetle, *Anophthalmus* ('blind') *hitleri*, dedicated to the Führer in 1933.

Naming species became a matter of good taste and refinement although Linnaeus himself had given obscene names to many sea squirts, and the erotic terms he used to describe the structure of flowers caused the *Encyclopaedia Britannica* to huff and puff: 'A man would not naturally expect to meet with disgusting strokes of obscenity in a new system of botany, but obscenity is the very basis of the Linnaean system.'

The respectable thing to do was to confer a descriptive term that characterised the creature and caught its distinctiveness. Some taxonomists had tried, but lacked the imagination. A common green seaweed of the shore is called *Cladophora* meaning 'branch-bearing', as if having branches set it apart from all the other plants. But other botanists vividly caught the look and feel of some seaweeds: *Calliblepharis*, 'beautiful eyelash'; *Petrocelis*, 'rock stain'; *Plumaria*, 'soft feathers'; *Chylocladia*, 'juicy branches'. Others descriptions were a little more puzzling, such as *Apoglossum*, 'sprung tongue', and *Helminthora*, 'worm sperm'. The zoologists got the hang of it too, with the narwhal *Monodon monoceros*, 'one tooth, one horn'; and the starfish, which although not fish were often star-shaped as their names imply: *Asterina gibbosa*, 'lumpy little star'; *Solaster papposus*, 'bristly sun star'; *Pisaster ochraceous*, 'lesser yellow star' (unfortunately, the taxonomist saw one of the occasional yellow ones although almost all of them are purple).

Mythological figures with aquatic connections abound in the names of marine animals like the snail *Triton*, 'Neptune's son'; and the worms *Amphitrite*, 'Neptune's wife'; *Nereis*, after the Nereid sea nymphs; and *Aphrodite* who rose from the waves – but which part of her anatomy is recalled by this plump, hairy, oval creature is unclear.

Sex is well represented in marine creatures. The name of the slipper limpet *Crepidula fornicata*, 'slipper arch', is supposedly because the shape of the shell reminded the taxonomist of the arches (*fornicari*) of the aqueducts in ancient Rome beneath which the brothels were situated – hence fornication. Perhaps he also noticed that these limpets climb up into high-rise stacks and those in the penthouse can inseminate even those three storeys below. Sometimes the biologist's imagination runs riot, as with the clam *Abra cadabra* or the curvaceous trilobite *Monroea*.

The reasoning behind a name can be difficult to work out. What, for example, is the connection between the gentle moss animal *Electra* and Agamemnon's daughter who murdered her homicidal mother, or the snail *Thaïs* and the Athenian courtesan who set fire to Alexander's palace?

Some names capture the spirit as well as the appearance of the animal. I warm to Gosse's *Lophohelia*, 'tuft of suns', for a shrubby, polyp-laden anemone. *Tyrannosaurus rex*, 'king tyrant lizard', could hardly be bettered for drama, whereas *Allosaurus*, 'other lizard', is the epitome of the unimaginative. A plant's perfume might also be noteworthy as in *Daphne odora*. Recently, a neighbour gave me a lovely garden plant. She called it a mirror bush, as it had dark super-glossy leaves, but there was no reference to mirrors in its Latin name, *Coprosma*, which means 'shit-smelling'. I am anxiously waiting for it to come into flower.

For many Victorian academics the careless mixing of Latin and Greek roots in the same name was considered a sure sign of an inadequate education. Others thought this was the last bastion of elitism. John Ruskin proposed that birds should be given simple, descriptive names in English, for soon children would use English 'for all scholarly purposes' and biologists clung to Greek and Latin simply 'to mystify the illiterate many of their own land'. Edward Lear lampooned 'scientific' names in his *Nonsense Botany* with inventions such as *Washtubbia circularis* and *Manypeeplia upsidownia*. In *The Water Babies* Charles Kingsley took a sideswipe at the biologists by giving a name thirty letters long to the aquatic sprite, explaining that 'They call everything by long names now, for they have used up all the short ones'.

There *was* a degree of snobbery about proper names, but English names are slippery and the same creature is likely to be called different things even in various regions of the country, or the same name appended to an entirely different species elsewhere. The North American robin may have a red breast, but is unrelated to the European robin. And of course we simply don't have an English name for any but the commoner species. The early naturalists tried to supply them but most never caught on. I like the curiously medical air of many of the names Gosse gave to anemones such as the 'wrinkled beadlet' and the 'red specked pimplet', and the old-curiosity-shop feel of Forbes's starfishes including the 'Shetland argus' and 'Wally Penson's seastar from Ireland'. Best of all are the old seaweed names that smell of the sea. It is not difficult to imagine a Victorian lady telling a bemused fisherman about the delights of tangle, dabberlocks and furbelows.

Few names succeed in giving both a description and a dedication. An American botanist, Sylvia Earle, discovered a tiny red alga in the shape of a parasol so she named it after her old professor, Harold Humm. She called it *Hummbrella*.

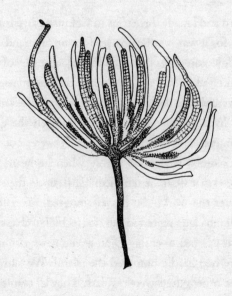

The Pleasure
of Piers

Isle of Colonsay, Hebrides

John Milburn and I made three trips to Colonsay, an island named after St Columba for it was here he first set foot on Scotland. An abbey was erected on Colonsay, and a priory on the adjacent islet of Oronsay. The islet was dedicated to St Oran (Odhran) who had himself buried alive for three days and then declared he had seen both Heaven and Hell and neither had lived up to expectations; in Heaven the good were not eternally content and in Hell the wicked were not unremittingly miserable. Columba recognised heresy when he heard it and immediately ordered that Oran be reburied: 'Earth upon the eye of Odhran, lest he corrupt the world.' So he was interred, never-to-be-retrieved. Perhaps Columba later regretted this, for he built a chapel over the spot – or maybe it was just to ensure Oran would never rise again.

Tourism had hardly bothered the island. We stayed at the only hotel and the other guests were regulars. They all seemed to be retired brigadiers. 'Been coming for twelve years now, me and the little lady', one confessed. 'Spot of fishing, birdwatching through the old binoculars, good chinwag over a G and T in the bar. What more could a chap want?' His plump wife sat quietly in a frock decorated with vines and extravagant chrysanthemums – lost in a jungle with no hope of rescue. She didn't say what a woman might want.

Like all the islands it is windy. For every gale that rattles the roof slates of Glasgow, seven shake Colonsay. On an otherwise calm day I saw a whirlwind waltzing across the fields sucking up leaves and fertiliser sacks, as if God were hoovering the front room. The hills grew nothing taller than bearberry, but in sheltered pockets on the leeward side were small reminders of the ancient Hebridean woods, copses of ash and oak with melancholy thistle and enchanter's night-shade sheltering beneath.

It was wonderful place for seabirds and it was here that I was first dive-bombed by terns. Since they swerve aside only an instant before they would hit your cranium, ornithologists advise you should hold a stick above your head. A lady writing to a Scottish paper was concerned that this might damage the tern and recommended the use of a stick of rhubarb instead.

We had come to study the pier. Britain had more piers than any other country in the world. Not mere jetties and landing stages, but seaside pleasure piers – 'works of enchantment hanging over the sea'. Historically, many harbours had a structure for fishing smacks to dock, but from the 1860s piers were built to attract boats laden with trippers from the cities. They alighted from the steamer on to what was effectively a stationary ship. Entrepreneurs soon realised that visitors would pay twopence, as Thackeray said, 'to go out to sea and pace this vast deck without need of a basin'. Piers became the rage and there was a rash of new protuberances sticking out over the briny, each longer than the last. At Herne Bay, Telford's pier stretched for 914 metres, later surpassed by the 2,158-metre construction at Clacton with its two-track electric railway. By the 1880s they were being widened to take ornate pavilions resembling potentates' palaces. Clacton's became a huge funfair on legs. Attractions included everything imaginable, from performing fleas to death-defying displays of diving 'professors' shackled in chains or enveloped in flames who plunged over the side into the ocean.

It was heady stuff when thrilling entertainment was combined with the wind in your hair and the giddy feeling of the waves rushing past beneath your feet. Best of all was the aphrodisiac effect that caused young women who were perfectly well behaved in London to 'give themselves up to a general *abandon* on piers which is astonishing'. The pleasure pier had well and truly arrived.

The pavilions often housed a theatre – some with thousands of seats. Now-forgotten stars shone in the pierrots until they were supplanted by the concert parties: the *Moonlight Follies* at Worthing, the *Fol-de-Rols* at Eastbourne and *Gay Cadets* at Bournemouth. Then there were the blacked-up minstrels who wished they were in Dixie, as did some of the audience. Gradually, this rich fare lost its charm and as pleasure piers became high-maintenance follies, they fell into neglect.

Although their ankles were forever in the water, the early wooden piers proved to be highly flammable, and those that the flames overlooked, the storms often destroyed. Even the later iron ones often had to be demolished when their spindly insect legs rusted and became unsafe. During the Second World War, several were breached in the middle to prevent enemy landings should there be an invasion. More than half of the hundred or so pleasure piers have now gone and inevitably more will follow.

The pier at the little port of Scalasaig on Colonsay is typical of the new generation of functional docking structures with no trace of pleasure pierdom. It stretches almost half a kilometre seaward to where the depth is still only three metres. It was constructed only a year before our first visit, so it provided an opportunity to follow the natural patterns of colonisation on artificial structures in the sea. We monitored what developed on the pier legs over their first five years.

As each pile had a well-lit outer surface and a shaded inner one, it was also an ideal place for studying the effects of underwater light. Although all the piles were almost identical replicates of each other, the outer surfaces on one side of the pier face north and those on the other side face south, so they would also show the influence of aspect.

After only a year patterns were apparent. Light-loving seaweeds clothed the outer surfaces whereas the shadowed inner sides were busy with barnacles, sea squirts, anemones and mussels. And everywhere there were eggs: the convoluted ribbons of sea lemons, clusters of the papery parcels of whelks and the dog winkle's elegant urns for embryos. This was colonisation in earnest.

Within three years the outer sides of the piles bore miniature forests of kelp with an undergrowth of leafy algae. These red seaweeds were shade lovers and when we removed the kelp canopy above them they bleached to a pale fluorescent orange within a day. The bulk of the kelp grew near to the surface then progressively diminished with depth, except for a sudden increase at the very base of the piles where there was a surprising increase in light because of reflection from the white sand beneath.

We tried to explain to a local what we were up to, but he was not convinced. 'I dinnae ken this at all – you get freezing cold in the sea yoo're saying, and after you've dragged up all this tangle you spend hours weighing and measuring it, and then you *throw it back*?'

'Well, yes.'

'Dinnae worry, laddie, yoor secret's safe with me.'

He was of course familiar with taking creatures from the sea for profit. When diving beyond the pier I came across a heap of pink litter. It was a pile of heads, thousands of them. They had until recently been important components of Dublin Bay prawns that had been snared in a fisherman's trawl. They should be difficult to catch as they live in muddy burrows beyond the reach of trawl nets, but fishermen soon learned that they pop out at dawn and dusk to feed and this is when they catch them.

Until the late 1950s they were just a nuisance, clogging nets cast in search of bottom-living fish, and were thrown back. Now they were beheaded, crumbed or battered, renamed scampi and were more valuable than fish. There were hordes of them around the Scottish coasts, far more than was ever realised, and in the Irish Sea the area trawled for them every year was twice the total area of the sea floor, so there must be many piles of detached heads on the bottom.

Just beyond the dead heads I saw the most gruesome sight I have ever encountered under water. Ahead in the murk was a fisherman in his hooded yellow smock, sitting up and swaying gently back and forth in the current. I was tempted to flee, but was hypnotised by the corpse. Then I saw it had no face. Had he, like the prawns, been decapitated? Slowly I realised that it had no head because it had no body. It was a discarded waterproof inhabited only by mud and water.

Its eerie animation came from the sway of the sea down there. Inside every wave as it rolls towards the shore, water molecules tumble in a circle. Deeper down, this orbital rotation gradually flattens out into an elliptical movement, which on the bottom becomes a to-and-fro swash. Every particle in the water, every scrap of loose weed drifted away and then returned. Anchored by a rock on the sandy bottom, a three-metre-long sugar kelp stretched out in the current then folded back over itself and gracefully unrolled until it was fully extended in the other direction like a dancer's arm. No sooner did it arrive than the flow reversed and the kelp slipped back again.

Inhabitants of the Hebrides have also learned to go with the flow. They have been given little choice. The history of the islands is the story of landlords and tenants, lairds and serfs.

Colonsay House nestles in an arbour of rhododendrons, magnolias and palms. It is the home of Lord Strathcona. Donald Smith, the first

Lord Strathcona, was an extraordinary entrepreneur and builder of the British Empire. He was chief executive of the Hudson Bay Company and became British High Commissioner for Canada. Smith was responsible for Canada's transcontinental railway, which transformed a country that was just two populated seaboards with nothing in between except thousands of kilometres of vacant prairie, tundra and mountains. Before the coming of the railway Canada was little more than 'a geographical expression'. When Smith died in 1914 he was the sixteenth richest man in Britain.

By 1967, the fourth Baron Strathcona was laird of Colonsay. He was married to Jane, the daughter of Earl Waldegrave. I never met her, but she almost gave me a criminal record.

We were diving under the pier, unaware that the royal yacht *Britannia* had anchored just offshore. Queen Elizabeth, the Queen Mother, had come to call on Lady Strathcona, whose sister was a lady in waiting to the Queen. To ensure that all the arrangements were satisfactory, the aide-de-camp sailed into the harbour in what might have been called an old motor boat had it not been adorned on either side with leaping golden dolphins.

Just as it approached the pier, I surfaced, grabbed a small black object with wires attached and went down again. The aide spotted me and almost fell overboard. When I resurfaced fifteen minutes later, I was apprehended by a naval officer and two stern fellows in plain clothes.

They stood very close to me and the ugly one asked the questions.

'Now, sonny, what exactly was that device you were fixing beneath the pier?'

'You mean the photometer for measuring light?'

'A photometer,' he repeated with a sneer, as if to imply that all saboteurs tried that one.

'I'm studying algae underwater, their growth, productivity and . . .'

By now his brain was filled with images of Semtex seaweed. As
soon as the Queen Mum stepped ashore on to seemingly innocuous
kelp – BOOM! He saw her feathery lilac hat frisbeeing into eternity.

Eventually they decided that no real assassin would make up such
a daft story.

When she trod ashore over the suspect seaweed I was not invited
to meet her.

Kelp and Selkies

Isle of Mull, Hebrides

At dusk I saw a lone man collecting seaweed cast ashore on Mull. Islanders had been gathering this free manure for centuries and it had played a major part in the fluctuating fortunes of all Hebridean islanders.

After the defeat at Culloden, most of the clan chiefs were stripped of their power. There was no one who felt a responsibility for the crofters, and the very idea of clanship was lost. Those in authority did not hold the islanders in high esteem. The sheriff of Inverness-shire described them as being 'infested with poverty and idleness', a view shared by Sir George Mackenzie of Coul who said, 'They live in the midst of smoke and filth. That is their choice. They will yet find themselves happier and more comfortable in the capacity of servants to substantial tenants.' The land was leased to lowlanders and Englishmen who were more concerned with profit than local prosperity. But crofting was subsistence farming of no benefit to a landlord and the islands were rich only in wind and waves that buried the beaches beneath heaps of cast-up kelp.

The word 'kelp' originally referred not to the big brown seaweeds, but the ash that remained when they were burned. Seaweeds could be incinerated to produce potash, which itself means 'ash containing potassium carbonate'. It was a vital ingredient of fertilisers and as Britain's agricultural economy grew, demand for potash outstripped supply. The Hebridean landlords were quick to see that they could provide the raw material.

Kelp gathering was labour-intensive, so in 1803 the landowners used all their clout to support an act of Parliament that they hoped would restrict emigration and retain their workforce. For almost sixty years the market for kelp remained buoyant and the population of the isles grew by 63 per cent. The rents were increased, so that to pay them tenants were forced to gather kelp as well as farm. A single laird, Lord Macdonald of the Isles, garnered £20,000 per year from kelp alone. Very little of this was invested locally; it went to meet the cost of high living in London.

It was too good to last. By the end of the Napoleonic Wars in 1815, duties on imported goods were eased and the kelp industry began to decline. By 1849, even the most profitable of the seaweed islands were impoverished and wretched. Those lairds whose wealth had come from weed were sinking into bankruptcy. Many sold their islands. Of the landlords who stayed, even the most humane saw their only salvation lay in sheep. The Blackface and the Great Cheviot that had turned the Highlands 'white with woollen snow' years before were now introduced to almost all the islands.

Sheep needed land and that meant the people had to go. Landlords who had once opposed emigration now encouraged it, often by force. The evictions on Mull were particularly severe and many tenants had to pay the cost of the writ against them.

The *Scotsman* considered that to be 'fruitful' when living in a barren country was 'proof of indolence' and the rising population must 'be removed from the vicious influence of idleness . . . walking about all day with their hands in their pockets, and at night sitting down and telling traditions about great lords and their mighty chiefs, and stories about ghosts and fairies'.

Country folk *must* be the custodians of the past, for city dwellers can be relied upon to forget. In Mull, I chatted to the old people who still nourished long-remembered tales, but they were full of sadness for those they had forgotten. 'If only I had listened to my grandfather when I was a child' was a common regret.

Almost all the old folk believed in fairies. No, more than that, they *knew* there were fairies. They had heard their seductive music, but hurried past because these were not just mischievous imps or gossamer-winged dryads, they could be ugly and malevolent.

Mull had long been known for wise women and hags, changelings and devils. It was here that the Hag of Winter fled each spring to expire. Mull was also a focus for fallen angels who plummeted into the sea with a hiss, for their veins ran with fire not blood. The local waters were also thick with mermen who came ashore, shed their tails and seduced the local girls.

Many folk believed in selkies, seals that lived as humans. Seals were widely feared and had long been considered half-human, a better percentage than many of us achieve. Although they had long ago forgotten how to speak, their plaintive moans and wails that emerged from the mist can still the heart.

In the old days when her clan set out on a seal hunt, one old woman was always seized with violent pains at the murder of her aquatic kin. Even hunters on their first kill felt uneasy when they looked into the solemn eyes of their victims, and heard the crying of the pups, so human you would swear they were babies. While some men clubbed seals to death, others crossed themselves if they came across a dead seal. There was consternation when infants were born with webbed fingers or toes – clearly the result of a tryst with a selkie while the husband was away. Selkies were unpredictable; they saved some men from drowning, but dragged others down to their death.

Throughout the Celtic world there are tales of a seal coming ashore and slipping out of its skin to reveal a woman. Sometimes a wily local hid her pelt so the maiden was unable to return to the sea and was condemned to remain as his captive wife. These are always stories of loss and longing; she spends every evening gazing out to sea and her relatives keep her captor awake with nocturnal laments for their lost sister. In the end, either she pines away or finds her hidden skin and returns to the ocean leaving her land-bound husband to die of a broken heart.

Throughout the eighteenth century, when seals came ashore to breed they were slaughtered for oil and meat (in spite of the smell), and to make sealskin boots and caps. 'When the crew are quietly landed . . . the signal for the general attacque is given from the boat so they beat them down with big staves . . . giving them many blows before they are killed. . . I was told also that 320 seals, young and old, have been killed at one time and place.'

With the starvation that followed the collapse of the kelp trade, seal killing revived out of necessity, but by the end of the nineteenth century it had become a pastime. Seals were forced to calve in ever more inhospitable places. By 1915, it was estimated that fewer than five hundred Atlantic seals were left in Scottish waters. An act of Parliament was passed to protect them from 'wanton destruction by increased visitation of our remote rocky shores of so-called "sportsmen" and thoughtless ruffians of all kinds'. In the twentieth century seals were shot because they competed with fishermen for fish, and a further Conservation of Seals Act offered protection, but allowed controlled culls.

Now, more than a third of the world's Atlantic seals are found in the waters of the Hebrides. They are also called grey seals although they vary from black to blotchy to foxy red. Their Latin name, *Halichoeris grypus* means 'Roman-nosed sea pig', which perfectly describes the big bulls that are two metres long and weigh three hundred kilograms. They come ashore in August to guard their pregnant harem and, with so many randy rivals around, dare not leave their post for six or eight weeks, not even to drink or feed. They may settle close to the sea or over four hundred metres inland and forty metres above high-tide mark. The females often fidget and bicker, but when calm they make soft haunting calls.

The pups are soon dropped and black-backed gulls fight over the afterbirths. The babies are suckled for only seventeen to eighteen days, but the milk contains almost fifteen times more fat than cow's milk so they put on 1.5kg per day and their mother loses about 4kg a day. After eighteen days the now almost skinny mother abandons her fat pup that must then fend for itself. Over half of them die within a year. Many succumb early on to the ravenous gulls, bites from rival cows, crushing by careless adults, and the predators that await them in the sea. The shore is littered with dead pups and they float in the putrid pools.

Before she leaves the beach, the female mates again with the bull, but the fertilised egg remains dormant for about three months before becoming implanted in the uterus when next she moults. Her gestation period lasts nine months, the same as for a woman, as befits a selkie, but delayed implantation means that she will return to the beach exactly one year later when the embryo is full-grown. In between, she will roam up to seven hundred kilometres away from the breeding site to feed and restore her weight. Over the years that she remains desirable to the male, she spends her entire time pregnant.

Seals seem to be as interested in us as we are in them. As we walked along the shore, their slick liquorice heads bobbed at the surface and turned to follow our progress. One dived and returned with a companion as if he had gone to fetch someone to verify what he had seen. They often accompanied us on our dives, swimming mostly at a distance so that we were always in view, but never close enough to be a threat. When I was snorkelling, a seal followed my every move, diving down with me and surfacing when I did. Only once did we meet a couple that were more interested in each other than in us. It was a young bull and cow who headed straight for each other, touched noses and then tumbled together in the surf, sometimes leaping clear of the water. Underwater they swam together; he approached, then she dived and he followed. They swam almost intertwined, lithe and liquid, tactile torpedoes, so engrossed that they never gave us a glance.

Perhaps it was a practice courtship or just exuberant play. I saw another seal playing frisbee with a jellyfish, repeatedly surfacing with it in its mouth and then with a flick sending it skimming into the air.

We also found them apparently dozing, sometimes head up above the surface, sometimes lying on the bottom. As we approached they just opened an eye, then swam lazily away.

Even wild seals quickly adapt to the presence of people. In 1961, a young Atlantic seal came ashore on the Dorset coast and after a week of eyeing a family on holiday, he allowed them to approach and stroke him. The next day he expected to be fondled. Soon he was resting his head in their lap or enticing them into the sea so they could swim together and play hide-and-seek. He embraced them with his flippers and nipped them gently with his mouth. When they left the beach for the day he followed as far as he could and wailed after them. On their return next morning he lumbered up to greet them. Then, after a long summer with 'friends', he returned to the sea for ever.

One young woman adopted an orphaned pup abandoned on the Hebrides and carried it halfway across Scotland in a tartan blanket.

Lora, the seal, soon adjusted to the menagerie of goats, rats, dogs, deer, birds and squirrels, plus a dog and a pony. The seal was fed on cow's milk enriched with oil and insisted on sitting on a lap to be fed. Lora soon learned to recognise thirty-five words, including the names of people and animals. She could recognise the postman from afar and waddled down the lane like an excited slug to receive the letters to bring home in her mouth.

She responded to music by almost going into a trance, a state that lasted for a while even after the music had ended. The exclamation 'Swim!' elicited the same response as 'walkies' to a dog. Lora was taken out regularly to the local loch to hunt for fish, but after she twice almost capsized the boat, she learned to sit patiently until told to leap overboard. Then, after seven years of living with humans, she went for a swim and was never seen again.

Even after the prolonged companionship of humans the call of the sea can be irresistible. I too know the lure of the ocean and the solitude of being underwater. Yet ironically, underwater is where you need a buddy most of all. In deep water off Mull, I turned for no particular reason to find that John was in trouble. He was unable to purge his flooded mouthpiece and was suffocating. He needed my air. It is one thing to practise giving away your mouthpiece to someone else in a swimming pool, but entirely different to do it thirty-five metres beneath the sea.

After taking a few breaths, John dutifully handed back my air supply. I took a couple of breaths and returned it to him. I clamped him to me with my legs so that me and my air supply couldn't depart in different directions. We inflated our life jackets and rose safely to the surface. John wasn't destined to drown in the dark waters of Mull.

He was an enthusiast who threw himself into whatever he did. On Colonsay we played table tennis in the hotel and it felt like the world championship. He loved new challenges. Years later he bought himself a micro-light aircraft for his birthday. On his first solo flight he lost control and plummeted to his death.

Lost Ships

Isle of Mull, Hebrides

It is not only marine biologists who become intoxicated by the sea. Nations viewed the oceans as a field on which to display their might and a battlefield on which to enforce their will. Warships have always been the largest fighting machines of their time. They were thought of as invincible – some were even called dreadnoughts.

But there *was* something to fear; the greatest enemy was always the sea. All their firepower was nothing compared to the power of the ocean. No one learned this lesson more cruelly than the Spanish.

The destruction of the Spanish Armada in 1588 did not take place at the hands of Sir Francis Drake in the English Channel, but far away on the lonely shores of Ireland and Scotland. Just as our weather had twice frustrated Julius Caesar's invasion in 55 BC, it also rescued England from the Spanish. The sea often plays havoc with history. Even a commemorative medal bore the inscription *Afflavit Deus et dissipantur*, 'God blew and they were scattered'.

After an eight-hour battle only three Spanish ships had been sunk and four run aground, but the Armada fled. The Spanish were too timid to risk running the gauntlet back through the Channel, so they sailed northwards then around Scotland to return south by the Atlantic. But whatever damage the English fleet could have inflicted, it would have been nothing compared to the ordeal that the waves and rocks had in store. In only two weeks twenty-seven galleons were wrecked on the Irish coast. In all, sixty-four ships were lost – half the original Armada.

Some of the ships were so damaged that their captains dare not risk them in the open Atlantic, so they sheltered among the Hebrides in neutral Scotland. One arrived 'beaten with shot and wether' in Tobermory Bay on Mull. Word reached Queen Elizabeth's Secretary of State that 'she is thought to be verie riche'. A week later she had been sabotaged and sunk.

It was so tempting to believe that the wreck at Tobermory was a treasure chest that fifteen generations of the Dukes of Argyll strove to retrieve it. The 'family wreck' almost wrecked the family. Over the succeeding centuries fifty expeditions sought the Tobermory treasure. Perhaps the main value of 'treasure' ships is for the employment of divers.

It is amazing that a vessel lying only forty-five metres from the town pier and just sixteen metres down should have yielded so little. The usual problem in treasure hunting is to find the wreck. At Tobermory the ship's position was known from the outset and as a result it was torn apart with primitive gear. By the time sophisticated underwater salvage equipment was developed, the wreck had been destroyed and was buried beneath tens of metres of silt.

It was the longest treasure hunt in history, yet for all this effort the total haul was the ship's bell, a few pewter candlesticks and plates, a silver spoon, an encrusted sword, two daggers and a pocketful of coins. I suspect that if the money consumed on all these expeditions had been spent at Sotheby's, it would have bought a roomful of treasures.

Far more treasure ships have sunk than ever sailed. Most lost fortunes were never lost in the first place, but as Joseph Conrad said,

'There are few things as powerful as treasure, once it fastens itself on the mind.'

It was not only the Armada that foundered on the west coast of Scotland. The Hebrides is where many ships came to premature deaths. There is hardly a headland or reef that isn't the headstone for a lost vessel and sometimes for its crew and passengers too. All the requirements for maritime disaster are to be found here: sharp-toothed skerries in narrow channels beset by fogs and sudden squalls, not to mention careless captains and drunken mates. One of the dispirited soldiers on an Armada galleon complained that 'there are fifteen types of weather in the Hebrides and fourteen of them obscure the lookout's view'.

You might think that an island would be reticent about its wrecks, for they hint at inhospitable coasts and sometimes the helpful offices of wreckers. But for the benefit of divers, Mull now boasts of the ships it has destroyed.

Two wrecks remain in my memory. Both were of ships that had sheltered from storms in the narrow Sound of Mull. The *Hispania* sounds as if it should be a Spanish galleon but it was a Swedish steamship that collided with Sgeir More (the 'big rock') in December 1954. She was mortally wounded and Captain Dahn ordered his crew to the lifeboats. Inexplicably, he refused to join them and when the ship sank an hour later he was seen standing at attention on the bridge saluting.

I visited the *Hispania* fourteen years or so after it sank. It sat silent and upright on the bottom thirty metres down as if at anchor. Indeed, its anchor chain still trailed from the bow. She loomed out of the haze, still intact, since only its cargo had been salvaged.

The *Hispania* was not a big ship, but its spars and masts seemed to soar towards the surface. Ships are amazingly tall. We are used to seeing only half their height for, like an iceberg, much is hidden below the surface. Ironically, it is only when the entire ship is submerged that its true height can be seen, although only by divers.

Just south of the *Hispania* lies the *Rondo*, which sailed under the flag of the Cunard company. In January 1935, she ran north in the teeth of a blizzard until forced to anchor for the night. But she dragged her moorings and drifted south for several kilometres until she collided with the islet of Dearg Sgeir. She perched there immovable while the salvors began to demolish her superstructure. Then a storm dislodged her and she slipped from the reef and dived into the depths. There she lay or rather, stood, for the *Rondo* lies along the length of the vertical cliff with her bow portion bent like a banana on the mud at the base of the cliff fifty metres down, while her stern is just below the surface. Her hull did a headstand. As all her top hamper has gone and even some of the upper hull plates were removed, she resembles a gigantic steel canoe, eighty metres long, filled with a jumble of pipes and derricks among its internal ribs. It is unimaginably strange to hover in the shallows looking down in the clear water with twenty metres of hull visible until it vanishes into the depths. No matter how long you stare, a vertical ship is difficult to comprehend.

On land, castles may crumble into romantic ruins, but if they remain reasonably intact we mend the roof then fill them with story-boards and exhibits so that paying customers can shuffle through over the ancient but newly swept floors. Sound effects and atmospheric lighting help to create an ambience of ancient times.

Wrecked ships need no such help. They ooze atmosphere and are the eeriest places on Earth. The surrounding haze creates mystery and a feeling of discovery. Sometimes snagged nets wreathe hulks in aquatic cobwebs that add an air of witchery. But it is the gloom inside that generates unease. It is impossible to venture into the black heart of a hulk without feeling that below in the darkness something awaits you.

They may also retain the feeling of sudden abandonment. In the cabins of the *Hispania* I found a cup with a broken handle, a scrubbing brush beside the bathtub and a lone shoe slowly filling with silt. Down there these mundane objects became imbued with a poignancy that they could never possess in a museum case. It never occurred to me that they were merely artefacts or souvenirs to be collected; they were still personal items belonging the crew.

Samuel Johnson declared that 'No man will be a sailor who has contrivance enough to get himself into a jail; for being in a ship is being in a jail with a chance of being drowned'. Wrecks are the result of a catastrophe and sometimes the victims remain within. The divers' nervous excitement is fed by the suspicion that the very next cabin they enter may contain the remains of a hapless mariner.

The unease is stoked by the knowledge that wrecks are dangerous places to be. It is easy to become disoriented in a confusion of corridors and decks even when a ship is well lit and afloat, but down here in the darkness amid clouds of silt, it is possible to get lost in the labyrinth where there is no easy escape to the surface. The tenuous artificiality of your existence is emphasised by the precious air expelled with every breath to accumulate as quicksilver pools on the ceiling. The trapped diver is unique among the condemned in that he can see his last breath.

It is easy to imagine that wrecks are dead buildings, becalmed and gently rusting. But they are far from being derelict. They have merely swapped the bustle of crew and passengers for the exertions of marine life. The rails of the *Hispania* were festooned with hydroids, and plumose anemones decorated the hull with a white-and-orange patch-work. The *Rondo*'s plates supported fields of dead man's fingers. All wrecks are havens for fish, but each may have its own characteristic fauna; when I was there cod and conger lingered in the dark holds of the *Hispania*, saithe and wrasse patrolled the open wreckage of the *Rondo*.

What they lack in variety of species, wrecks make up for in the startling abundance of their inhabitants. I have seen them entirely furred with moss animals, spectacular gardens of feather stars or clusters of transparent sea squirts like posies of condoms. Many hulks have cellars tenanted by sharks.

There are 45,000 known shipwrecks around the British Isles and perhaps hundreds of thousands still undiscovered. They provide the accommodation that many marine creatures crave and usually all the rooms are taken. Other man-made structures such as oil and gas rigs attract a fauna otherwise absent from the adjacent plains of sand and mud. All over the world, retired barges, tugs, destroyers, cars, railway coaches and trams have been scuttled to make havens for sporting fish. Now shredded tyres, concrete or compacted fuel ash have been moulded into interesting shapes and submerged to make artificial reefs. Some are even constructed from the cemented ashes of the cremated, but still environmentally-minded, dead.

Home Deep
Home

Argyll, Scotland

In 1967, I was invited to take part in an underwater experiment called the 'Kraken', to man a submerged laboratory twenty-seven metres down. It was stimulated by the diver's frustration that the time spent working underwater was limited by the need to return safely to the surface. At twenty-seven metres the pressure is almost four times that at the surface and breathing compressed air causes problems. Air is 78 per cent nitrogen, which at depth causes an intoxication impairing the diver's judgement and making accidents more likely. Nitrogen is readily taken up into the body's tissues. The deeper you go and the longer you stay, the more nitrogen is absorbed, which becomes a problem when you begin to return to the surface. Ascending reduces the pressure and the nitrogen comes out of solution and expands into bubbles in the blood and joints – the dreaded bends. Divers who venture deep must spend an inordinate amount of time halting during their ascent to allow the nitrogen to be respired away safely. Spending only an hour ninety metres down requires a decompression period of seven and a half hours.

Although nitrogen uptake increases with the time spent underwater, there comes a point at which the body is saturated with it and unless the diver ventures deeper, the burden of nitrogen doesn't increase. Nor does the time required for safe decompression. It follows that instead of making repeated short dives to the sea floor, it should be more efficient to spend days or even weeks underwater and then have a single period of decompression at the end of your stay. This is called saturation diving.

The idea of having a prolonged stay on the bottom of the sea is not new. In 1774, a bold English carpenter called John Day boasted that he was to spend twenty-four hours beneath the sea. He constructed a sturdy wooden hut and lashed it to the deck of an old sloop, the *Maria*. With Day inside, the boat was scuttled and sank forty metres into Plymouth Sound. The next day he was to release the shackles and bob up to the surface to public acclaim. It was the very first saturation dive, and the longest. He's still down there.

The idea of saturation diving was first suggested in 1942, but it was not until the late fifties that an American family doctor called George Bond reiterated the principle and was seconded to the US Navy submarine laboratory. After six years of experimentation on animals he subjected human volunteers to simulated saturation dives in pressure chambers, replacing the nitrogen in air with helium to reduce the likelihood of both intoxication and the bends. Bond now felt confident to run a trial under the sea, but the cautious navy bureaucracy demanded more tests. Thus Bond, who had done all the groundwork for saturation diving, was not the first to demonstrate its capabilities in the sea. That fell to Edwin Link.

Ed Link was a modest, energetic man who had made his fortune in the 1940s by inventing the very first flight simulator for training pilots on the ground. The simulator was described as the result of 'a collision between the minds of Walt Disney and Heath Robinson', but it worked. He remained an inventor and, having read Bond's accounts of saturation diving, built a metal cylinder three metres long and less than a metre wide in which he sealed himself for eight hours at a depth of eighteen metres, then spent six hours decompressing. He called it the 'Man-in-the-Sea Project'.

Link's other interest was underwater archaeology and this had introduced him to a young diver called Robert Stenuit. Next thing he knew, Stenuit was spending over twenty-four hours sixty metres down in the cylinder. In deeper and longer trials he resided in a rubberised inflated 'tent' with an open hatch in the floor, only the pressure of the gas inside keeping the ocean at bay. The sealed cylinder was used as an elevator to bring him back to the surface still under pressure. It was then laid on its side while Stenuit decompressed inside. Well into the long decompression schedule, he suffered pains in his wrist and for fear of the bends he was repressurised and the whole process had to begin again. He was incarcerated in this steel coffin for over sixty-five hours – not an experience for the claustrophobe.

Jacques Cousteau was soon in on the act and built a multi-winged 'Starfish house' with proper bunkrooms, a laboratory and a changing room with access to the sea via an open hatch. The house boasted a few home comforts such as a sunbed, a cordon bleu chef and a couple of pet parrots called Claude and Armand. Once the five aquanauts were installed, everything, including the parrots, had to be brought down sealed in a pressure cooker.

The Kraken laboratory I was slated to inhabit would be a demonstration of how much more biological research could be achieved by living underwater than by daily dives from the surface. However, just as the original giant kraken was elusive, so too was the Kraken lab. It was budgeted to cost a modest £15,000 to £40,000, but even that proved too much to raise and the project was shelved. Perhaps it was just as well, for the plans were both vague and frightening. If power could not be supplied from the surface it 'could be replaced by a generator of some sort, probably nuclear'.

Within a couple of years other biologists did indeed work and sleep beneath the waves but not without incident. The US Navy devoted huge resources to their Sealab project off California, yet the Sealab house had an electrical fire and the aquanauts had to bail out. They had not realised that they risked incineration while surrounded by ocean – the world's largest fire extinguisher. A diver died and the programme was abandoned. Perhaps I was lucky not to end up drowned or hypothermic with only a nuclear generator to keep me warm. There are enough unsuspected dangers below.

One underwater house for biologists did prosper. The Tektite habitat was the ugliest house ever built and resembled a pair of gigantic waste-disposal units. It was manned for the first time as Neil Armstrong stepped on the moon. Indeed, it was part funded by NASA who wanted to study the psychology of people confined under stressful conditions. The male inhabitants received some publicity because America was simultaneously exploring both outer and inner space, but the next year, 1970, the media suddenly took an avid interest. It was the fault of a petite marine botanist called Sylvia Earle. She applied to join the relay of biologists who inhabited Tektite during the summer, but the planners baulked at the very idea that men and women in damp proximity should bed down together, even if it were only the seabed. Aquanauts had always been men; even Cousteau's parrots had been male. The organisers came up with a novel solution. How about an all-female team with Sylvia in charge? The *Boston Globe* headlined the news: HOUSEWIFE TO HEAD TEAM OF FEMALE AQUANAUTS.

So Sylvia, 'Her Royal Deepness', was off to the US Virgin Islands with four others – not aquanauts but 'aquanaughties', 'aquabelles' and 'aquababes'. They spent two weeks fifteen metres down and when they rose from the waves they were briefly more famous than Botticelli's Venus.

By 1978 there were perhaps fifty underwater habitats in the

world's oceans, but almost all of them lasted for only one or two seasons before they were abandoned. Cousteau's Starfish house now quietly corrodes on the bottom of the Red Sea.

Link, Cousteau and many others had believed that before the end of the twentieth century the sea floor of the continental shelf would be littered with pressurised bungalows from which residents would emerge to excavate wrecks, or scoop up diamonds and manganese nodules. For relaxation, they would weed allotments of sea vegetables or ranch fish. But Wetropolis never came to pass.

To sustain ourselves underwater is an unwieldy and costly business requiring a large logistic support from the surface. Saturation diving *is* now widely used in deep-sea operations, but only when immense profits are likely. Even so, the divers don't live on the sea bottom. For the duration of the task, sometimes six weeks or so, they inhabit comfortable pressurised cabins on board the mother ship. Each day they descend to their workplace in sealed chambers, slip out to do the job then ascend again under pressure. It would make no sense to have the people stationed far below in remote submerged habitats where they are inaccessible should anything go wrong. On board the ship, doctors can pass through an air lock to attend to them.

At best, underwater habitats were spartan. In the warm waters of the Red Sea Cousteau's aquanauts baked and 'fountained perspiration' whereas off France they shivered in the submarine chill. In both places the 100 per cent humidity was almost unbearable. When they breathed a helium mix instead of nitrogen they spoke in the irritating and incomprehensible brogue of Donald Duck. They became tired and after a week or two their sense of time and taste and smell faded. With schools of comestibles swimming all around them, Earle's companions confessed that the only fish they consumed were in frozen sticks from Safeway.

Then there is perhaps the most disturbing thing about saturation living. The surface is no longer the route to safety. On the contrary, it is where the greatest danger lies. To rise precipitously would be to suffer paralysis and death. While working away from the habitat, the regulator on Sylvia Earle's aqualung failed and her air supply ceased. She was only twenty metres from the surface; it would have been easy to rise and breathe fresh air and see the sky, but perhaps for the very last time. The house and safety were three hundred metres away. Fortunately, her companion realised she was in difficulties and offered her mouthpiece. Sharing air, they swam back to the Tektite house. It must have seemed a long journey.

Once there were dozens of underwater laboratories manned by diving scientists, now there is only one. It sits alone like a shed at the bottom of a damp garden.

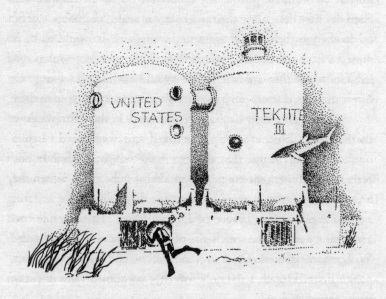

Just
Visiting

Nowhere

I have rarely visited a coast without at least snorkelling to view the shallows. It is an ancient tradition. Long before history began, naked divers plunged into the Pacific and the Mediterranean to find the living riches of the sea. Their feats were first recorded in Mesopotamia as early as 4500 BC. The early skin divers searched for sponges, coral and clams lined with mother-of-pearl. Chinese divers supplied their emperor with pearls as early as 2250 BC and divers in the Persian Gulf were said to hold their breath for several minutes and fill baskets full of pearls from eighteen metres down. Divers were later busy in ancient Assyria, Korea and India. In the late thirteenth century Marco Polo observed pearl divers in the shark-infested strait between Sri Lanka and India during the open season from April to May. Thereafter, the divers sailed 480 kilometres to fish another ground from September to October.

In ancient Greece, divers held their breath and descended to almost thirty metres in search of sponges. Glaucus, who built the *Argo* in which Jason voyaged in search of the Golden Fleece, was a sponge diver. He ate a magic seaweed that allowed him to stay down longer. Perhaps he consumed too much, for one day he failed to return. The ancient Greeks padded their helmets with sponge and Roman soldiers carried a sponge to dip into a stream to drink. They also used a sponge to wash their bottom. Ideally, they had *two* sponges.

Amazingly, commercial skin diving without the help of breathing equipment continues to the present day. Divers in southern Korea and Japan known as Ama can hold their breath for up to two minutes in search of clams and edible seaweed. Ama in Japanese is written as two characters meaning sea-man, but it's the women who dive. Many years ago men went under too, though most refrained – they claimed it made them sterile. The Ama women have a different view. They say it's because women are calmer, can hold their breath better and can tolerate the cold water three or four times longer. A village elder attributed this to them being 'covered with fat like seals'. Not so. The sienna-skinned Ama divers have athletic, feminine figures.

The Ama aren't born to dive like seals, they have to learn how to do it. When eleven or twelve years old they practice in the shallows and increase their lung capacity. Even pregnancy is no impediment and they often work right up to the day of delivery. A few continue diving into their sixties or, in one case, her eighties, but they say that few live past sixty, perhaps because of the hard breathing and cold water.

Hekura Jima is a treeless clump of black cliffs only twelve metres high and encircled by the ocean thirty-two kilometres off the west coast of Honshu. Legend has it that even the rocks are merely petrified waves. It is inhabited largely by gangs of feral cats, but for three months each summer the Ama came to collect abalone shells full of meat and lined with mother-of-pearl. A schoolmaster, doctor and Buddhist priest came too. There is a shrine on the island dedicated to the 'Princess of the Hidden Bay' who inhabits many legends of the sea.

The Ama dives almost naked apart from a thong and headscarf. The skin on her face is protected by white cream and the eyes are covered by a pair of swimming goggles flanked by two small, inflated bladders. As she descends, the air in the bladders is squeezed into the goggles to prevent the rims pressing into the eye sockets and forcing out the eyeballs like squeezed grapes.

Usually she works from a boat, assisted by a boy or her husband. She plummets to the bottom, helped by a weight belt or clutching a boulder, then searches the sea floor, her body often vanishing into the waving meadows of seagrass. After forty-five seconds or so the helper hauls her up with the line fastened around her waist. She rests briefly on the surface while he retrieves the boulder, which he has also tied to a rope. He hands her the rock and down she goes again. The Ama collects eighteen to twenty-four metres beneath the waves, making thirty or more dives an hour. After each hour she rests a while in the boat. This routine is repeated all day long.

Occasionally, professional skin divers accomplish exceptional feats. In 1913, a Greek sponge diver called Stotti Georghios was said to have dived down sixty metres without equipment to attach a line to the anchor of a sunken battleship. He took four minutes to complete the job. It was inevitable that sooner or later someone would see this as a sport.

Free diving is the art of going deeper and staying longer underwater without the benefit of breathing apparatus. As in any sport there are several disciplines. Some events simply require the contestants to hold their breath for over eight minutes to approach the record. In the more active events they must haul themselves down a vertical line and then (ideally) return to the surface without the help of anything other than a bathing costume. The record is currently eighty metres. When a vertical sled is used to accelerate the descent and a large balloon to aid the ascent, the record is an astounding 154 metres.

Free diving began in 1949 when a Hungarian called Raimondo Bucher won a bet that he could swim down to thirty metres wearing just a mask and fins. By the 1960s the sport was thriving and Frenchman Jacques Mayol, who had seen his father drown while diving for sponges, became the first man to reach the hundred-metre mark – at the age of fifty-six. His rivalry with the Italian Enzo Maiorca (who had been the first to reach fifty metres) was fictionalised in Luc Besson's cult movie *The Big Blue*. Mayol and Maiorca were friends as well as rivals, but today's heroes, the legendary Italian Umberto Pelizzari and the 'pugnacious' Cuban Pipin Fererras, are so bitterly at odds, they even refuse to appear together at the same events. In 1999, Pelizzari shattered the 'No Limits' record by reaching 150 metres and flouting the tacit agreement to break records by just a metre or so each time to give everyone a shot at glory and maintain the interest of the public and sponsors.

Tanya Streeter is a female free diver who looks as you imagine a mermaid should. But she dives like Superwoman. In July 2003, she needed only three and a half minutes to extend one of the *men's* records by seven metres to a staggering 122 metres.

Free divers descend into a blue nothingness simply because it is there. As far as the ocean is concerned, they are only visiting. They are not merely reaching down into the sea, but down inside themselves. Either way, as John Donne wrote four hundred years ago: 'Sea discoverers to new worlds have gone.'

Attempts to return to the sea are nothing new. Whales rediscovered their aquatic roots 35 million years ago. The danger in such a radical transition from a land- bound ancestor is that the intermediate stages may not be well adapted to either milieu and therefore vulnerable to both terrestrial and aquatic predators. On the other hand, this doesn't seem to inhibit modern penguins or seals that must breed on land where they are clearly less agile than in the water. Indeed, it may be an advantage to be able to pop into the sea when a bear lumbers by and get out when a shark appears.

Marine mammals seem to have come into their own as replacements for the giant marine reptiles such as the ichthyosaurs and plesiosaurs. When these became extinct it left an empty niche that mammals could fill without fear of competition. Some still bear reminders of their ancestry. The fossil record reveals that dolphins were originally derived from small shrew-like animals whose locomotion was a lolloping scamper in which the back flexed vertically. Hence the dolphin's tail beats up and down not side to side like that of a fish or reptile.

Most land animals can swim when the need arises. The record swim for an elephant is forty-eight kilometres, and they even go snorkelling. Aristotle observed that 'Just as divers are sometimes provided with instruments through which they can draw air from above the water, and thus remain for a long time under the sea, so also have elephants been furnished by nature with their lengthened nostril, and whenever they have to pass through water they lift the nostril above the surface'. The pressure on the submerged chest of a snorkelling elephant is 20 per cent greater than that of the air in its lungs, enough to burst all the fine blood vessels in the membrane that encloses the pleural cavity, a fluid-filled bag allowing the lungs to inflate uniformly within the irregular cavity of the chest. Bursting these capillaries is potentially fatal. But uniquely among mammals elephants have dispensed with a pleural cavity and instead have layers of connective tissue lacking fragile capillaries. All this padding allows the lungs to expand safely against the chest wall when the elephant goes snorkelling.

The need to breathe air is clearly not a major impediment to living an aquatic life. Of course, it helps if you can hold your breath for well over an hour like the sperm whale and the bigger seals. It was thought that the Fitzroy turtle could hold its breath for three days, until it was caught cheating by breathing through its bum. Diving animals are so adept at holding their breath that they do it all the time even if there is no reason to. A seal dozing on the beach often stops breathing for long periods.

With such prodigious breath-holding abilities it is no wonder the leatherback turtle can descend to 640 metres, the elephant seal and many whales dive down over a kilometre and the sperm whale ventures well over two kilometres beneath the sea. Even the emperor penguin, which holds it breath for only ten minutes or so, has been seen five hundred metres down.

Of course, they have learned a few tricks to conserve energy and therefore oxygen. Seals, dolphins and whales have been observed to begin their descent with just a few powerful thrusts of their flippers then to sink effortlessly. Guillemots are accomplished underwater swimmers and have been filmed playing around at a depth of ninety metres, but they begin a dive by plunging into the sea from a great height so that their momentum carries them way down before they need to swim. It is far easier to reduce the effort of diving than to increase your oxygen supply.

The sea is frequently cold and water conducts warmth away from the body twenty-five times faster than air, so insulation is essential. If you fall overboard from a ship you will probably die from hypothermia. When the *Titanic* went down everyone in the lifeboats survived, but all those in the water died although it was flat calm and they were wearing life jackets. Survivors said that all cries from those in the water ceased within forty minutes. Even when the *Laconia* sank in the warm Mediterranean in summer two hundred passengers floating on the surface died from hypothermia.

How then do Ama divers cope? Humans respond to chill by shivering – an involuntary exercise that generates heat. Scientists kept Ama divers in tanks of iced water for up to three hours and found they could tolerate colder temperatures than non-divers without shivering. It was assumed that the Ama had been 'hardened-off' by their training to suppress the shivering response which increases heat loss.

Amas subject themselves to a greater cold stress than any other group of human beings. They lose around six hundred kilocalories during a single diving shift in winter, and their deep body temperature drops by well over two degrees centigrade. Half the Ama's fat is lost in winter as she lives off her reserves.

Diving marine mammals never skimp on insulation. There is no such thing as a skinny dugong or an anorexic whale. The blue whale is lined with 170 tons of blubber, almost half its body weight. Seals may be heavy waddlers on land, but thanks to blubber they are sleek, streamlined and snug underwater The trouble is that as fat is less dense than water it adds buoyancy. Sinking is easier if you are denser than water. So most marine mammals don't take the biggest breath possible before diving, they breathe *out* and half empty their lungs, presumably to reduce their buoyancy. Ama divers do the same and dive with their lungs only 85 per cent full.

But how do air breathers get enough oxygen from a single breath to sustain them on a long dive? One method is to hyperventilate – take several deep breaths just before diving. This not only sucks in more fresh air, it also flushes out all the stale carbon dioxide lurking in the recesses of the lungs. It can prolong the duration of a breath, but the problem is that the stimulus to breathe comes not from a lack of oxygen, but from a build-up of carbon dioxide. If it is artificially depressed, the diver has no desire to breathe even when his oxygen levels are dangerously low. Many a skin diver has died after hyperventilating. Ama divers hyperventilate briefly before diving, but often make a loud whistle every time they blow out. They say it makes them feel better and protects the lungs, but perhaps it prevents excessive hyperventilation.

Diving mammals store lots of oxygen much as a desert cactus stores water. They keep it mostly in their blood and muscle. These are big reservoirs, for an elephant seal has three times more blood than we do and ten times more oxygen-holding myoglobin in the muscle. Meat is flushed pink with oxygen-rich myoglobin, the muscles of diving mammals are almost black with it.

It is well known that when diving animals are underwater their blood supply is allocated to the most oxygen-sensitive organs such as the heart and brain, which cannot function without it. This is at the expense of all other tissues, including the muscles. They must get along without oxygen, which is tolerable in the short-term, until, as in athletes, the build-up of lactic acid impairs their function. The so-called 'diving reflex' also allows the heart to slow down, for now it is supplying a much smaller circuit. All this reduction in physiological activity cuts oxygen consumption and prolongs a dive, especially the occasional deep excursion, but it is not a strategy that can be used repeatedly by accomplished divers such as sperm whales that spend 90 per cent of their time at depth yet never surface out of breath.

We seem to retain some remnant of the diving reflex. Studies of free divers and Ama women have shown that if we merely hold our breath our internal engine continues to turn at the same rate, but if our face is in cold water the heart rate slows by 40 per cent and the blood supply is diverted to the vital organs. The Ama's basic metabolic rate is also elevated in winter by about a quarter, just as it is for Inuits enduring the Arctic winter.

Sadly, even all the Ama's interesting adaptations may not be enough to save her. In the early 1960s there were 30,000 Ama divers in Korea and Japan but by the 1970s there were only 7,000 left in Japan. They now use fins and rubber suits, but soon they will be extinct.

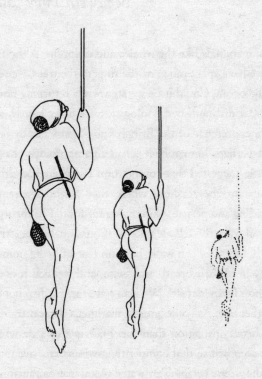

Wet Behind
the Years?

Somewhere in the past

We too are mammals like the whales and dolphins. Although in our evolution we have spent much of the time in the trees, we seem to be drawn to the ocean. Could it be we are merely returning home?

In 1960 a distinguished zoologist, Sir Alister Hardy, spiced up his talk at a conference of the British Sub-Aqua Club by outlining a notion that perhaps human beings had an amphibious phase in their evolution. He suggested that competition elsewhere had driven some early humans to the seashore where they adopted a semi-aquatic lifestyle, wading and perhaps skin diving for food. His theory was that our body still wore the telltale tokens of our aquatic ancestry. Babies, he argued, before they can walk, have no fear of being immersed and instinctively hold their breath and swim, and webbed toes or fingers are a common birth 'defect'. We also have far less fur, but more fat, than any other ape. As some aquatic mammals demonstrate, in water blubber is better insulation than fur. Hair is also a drawback when swimming, so much so that competitive swimmers shave their bodies to reduce drag. But foraging in water requires endurance not speed, and Arctic swimmers and divers such as polar bears and elephant seals are *more* hairy than their closest relatives.

Most curious of all, Hardy claimed, the down that covers our skin is arranged in a distinctive pattern that matches the hydrodynamic flow of water over the body when swimming. His account of hairodynamics is almost true, but on the shoulders, the leading edge of the body that meets the flow, the hairs point upwards (i.e. forwards), the 'wrong' direction. Apart from the groin and armpits where hair growth is related to the dispersal of scents from our glands, it grows in greatest profusion on the top of the head — we're not furred, but we are thatched. Hardy concluded that being naked was an adaptation to an aquatic life long ago, and the hair on our head protected the only part of the body that was habitually out of the water.

He quickly regretted the speech, for it was widely reported in the newspapers, and ridiculed. Professional anthropologists rejected it out of hand or simply ignored it. It is, of course, typical for any group of professionals to close ranks and snipe at interlopers. Yet in those days the proposed schemes of human evolution were based on a few fragments of skull here, a bit of leg bone there. 'The timing is wrong,' they said. Yet anthropologists were always having to revise their views because the latest find revealed that something must have happened much earlier than anyone had imagined or more recently than anybody had believed. It is only very recently that we have discovered that the Neanderthals and *Homo sapiens* are much less closely related than previously thought. Indeed, we are not even partially descended from them. We may never know for sure when *Homo sapiens* lost his fur.

The problem was that anthropologists *knew* that water played no part in human evolution. *Homo sapiens* had come down from the trees on to the plains, stood erect in the sunlight and lost his pelt because it was unbearably hot trotting round the tropical savannah in a fur coat. Surprising then that the camel and the cheetah manage so well.

Anthropologists knew the conditions in which early humans lived because their ancient remains were invariable found on hot dry plains. It never occurred to them that our physiology is reliant on frequent access to fresh water. We can't last long without a drink and waste profligate amounts of sweat and urine for a savannah dweller. Nor did they check what the climate was like when the fossil bones were still running around. It was not until the 1990s that climatic recorders such as ancient pollen grains were examined, and in site after site in Africa it became clear that early man lived mostly in forests, some with lianas clinging to the trees. There wasn't a savannah to be seen. Seventy years after the 'savannah hypothesis' became dogma, one of its leading proponents declared, 'We were all profoundly and unutterably wrong!' But this does not mean that Hardy was right.

It is not clear when humans first stood erect or even that the early progenitors of humans were lumbering semi-simian hulks. A recently discovered fossil hand 3.3 million years old came from an ape that did *not* walk on its knuckles. Almost all vertebrates walk and run (usually faster than we do) on all fours. Apes and monkeys take the weight mostly on their hind legs, and many of them can stand upright, sometimes for long periods, and even walk a short distance on two legs. Gibbons often stroll along branches using two legs, with a hand trailing on a parallel branch above. Chimpanzees can run quite rapidly on two legs for a short distance. So even in our ancient tree times we were probably accustomed to walking upright on occasion. Therefore opting for two legs most of the time was not a giant leap for mankind. The mystery is *why* we exchanged the economy, stability and speed of four legs for the inefficient bipedal teetering that we call walking. The suggestion by those who propound an aquatic ancestry for humans is that wading for seafood supplied the need to stand upright as well as the necessary support for learning to walk. But it doesn't hold water. I frequently wade in the shallows to search for specimens, not by standing upright, but by stooping forward so that I can see what's on the bottom. And anyone who thinks that it is easier to balance or walk when wading in the sea has never done it.

We have been called the naked ape and the reason for hair loss in humans is probably sexual. Today, the majority of men prefer hairless women. Females are more tolerant of hirsute men, but for many women hairy backs are several tufts too close to the gorilla. Smooth bare skin reveals a healthy body and would enhance selection by a potential mate. It can also transmit sexual signals such as blushing. Dense hair not only hides the view, it is also a nest for numerous parasites, some of which cause disease. Hairless skin does not need the incessant grooming that occupies half the chimpanzee's day, leaving early humans far more time to find food or shelter.

Without the insulation that hair provides, early humans might have frozen during the chill of the night had they not had a layer of fat just beneath the skin. We are the plumpest primates and the only ones with chubby babies. Apes only become obese in captivity. Although deep-diving vertebrates rely on blubber (up to fifty centimetres thick in whales) to keep them warm, the thin fat layer coating an athletic person is hopelessly inadequate insulation in water, hence our susceptibility to hypothermia.

Our fat does, however, provide some buoyancy when we go swimming. Artefacts found in southern Spain that are well over a million years old indicate that early man, *Homo erectus*, took a short cut in his migration from Africa by swimming the five-kilometre Strait of Gibraltar.

Much has been made of the fact that like diving mammals we too possess the diving reflex in which the heart slows and blood is directed only to the vital organs. But in humans this is merely a token response. In experienced skin divers the heart rate is only reduced by about 40 per cent, (though better than the meagre 20 per cent reduction shown by tyros), but experiments on divers holding their breath and swimming demonstrate that this reflex is too weak to reduce the body's oxygen consumption. In any case, it takes about forty seconds for the full effect to develop, by which time the average breath-holding dive would be over. If the diver breathes in before diving, it slows the development of the effect even more.

Language is perhaps the greatest difference between apes and us. They can of course communicate, but only we have the ability to utter long and subtly modulated statements. Although the structure of our larynx is very similar to theirs, it has slipped down the throat below the back of the tongue. This means that unlike most vertebrates we can't drink and breathe at the same time. Thus we have long practised holding our breath, for we must do so every time we swallow, otherwise it 'goes down the wrong way' and we choke. For apes and all terrestrial mammals breathing is largely involuntary, but humans exercise conscious control of breathing and, like diving mammals, can hold their breath when going underwater.

However, if we were descended from a truly aquatic ape, why are we not endowed with the essential attributes of a diving animal such as a waterproof skin that doesn't go wrinkly after ten minutes in the bath, effective insulation, a body with a large mass and small surface area, and eyes that can see underwater? When athletic we are too lean, and our extravagantly large surface area is because we are all arms and legs, ears, fingers and toes. Of course, our hands are manipulative wonders, but in water the hand can either reduce heat loss or retain its dexterity, not both. As for our eyes, they are 'designed' to work in air. Unlike air, water has a similar refractive index to that of the eye so the lens can't bend the light sufficiently to focus it on the retina. Thus underwater our vision is hopelessly blurred.

But even if our original ape ancestors were never truly aquatic, many early humans were indeed amphibious, and this may have had major consequences for our evolution. The human brain showed a sudden growth spurt during the last 200,000 years, with a 50 per cent increase in cranial capacity from *Homo erectus* to *Homo sapiens*. Two types of fatty acids are responsible for brain growth. Both are slow to form in the body and hard to come by on land, but are abundant in aquatic creatures inhabiting rivers, lakes and the ocean. Many ancient human settlements are beside the sea (though not until much later beside lakes) and can be recognised by the huge shell middens, the remnants of thousands of meals. Coastal sites in South Africa and on the Red Sea reveal that 100,000 years ago the main food of *Homo sapiens* was shellfish. This was a recipe for brain expansion. Even my granny knew it: 'Fish is good for you,' she would say. 'It's brain food.'

Anthropologists are at last coming round to the realisation that many of the obvious differences between humans and chimpanzees, with whom we share over 99 per cent of our genes, are explained by a waterside past in which humans went wading and diving and by 45,000 years ago, were sailing on rafts. We were fishers and sailors long before we were farmers and this phase lasted a long time. Neolithic Britons underwent a sudden shift from a marine-based diet to terrestrial food only about 5,000 years ago.

It made good sense to be coast dwellers. Living inland required a large area to support a family of hunter-gatherers, but collecting from the shore and shallow sea offered much richer pickings. Coastal living also determined the direction of their migrations. If local resources were depleted, they would not move inland or out to sea, but along the coast, and by moving at the rate of one mile every year, in only 12,000 years they could migrate from Europe to Asia.

Humans need few adaptations for a life of scavenging on the shore, wading and perhaps duck diving for food. Several species of monkey can swim and macaques even dive. There are some modern communities that still live an amphibious existence in which children are taken to the water almost from birth. Herman Melville saw an example of this in the South Pacific when strolling by a river:

> I observed a woman sitting upon a rock in the midst of the current, and watching with the liveliest interest the gambols of something, which at first I took to be an uncommonly large species of frog . . . I could hardly credit the evidence of my senses when I beheld a little infant, the period of whose birth could not be extended back many days, paddling about as if it had just risen to the surface, after being hatched into existence at the bottom. Occasionally the delighted parent reached out her hands towards it, when the little thing, uttering a faint cry and striking out its tiny limbs would sidle for the rock and be clasped to its mother's bosom. This was repeated again and again, the baby remaining in the stream for about a minute at a time. Once or twice it made wry faces at swallowing a mouthful of water, and choked and spluttered as if on the point of strangling. At such times, however, the mother snatched it up and by a process scarcely to be mentioned obliged it to eject the fluid.

He witnessed the same performance every morning and evening:

> No wonder the South Sea Islanders are an amphibious race when they are thus launched into the water as soon as they see the light. I am convinced that it is as natural for a human being to swim as it is for a duck, and yet in civilized

communities how many able-bodied individuals die like so many drowning kittens from the occurrence of the most trivial accidents?

Worlds Within Whorls

Isle of Jura, Hebrides

The telephone rang. 'I run the hotel at Craighouse on Jura,' said a man's voice. 'I'm thinking of organising activity weeks for guests with an expert on hand to guide them. I thought they might enjoy a week of marine biology, searching the rock pools, that sort of thing. Do you know anyone who'd be willing to help?'

I certainly did.

The Hebrides are islands of ambiguity, with the Atlantic forever pounding their western shores yet surrounding them with balmy waters so that while Edinburgh shivers in winter, their climate remains frost-free and mild. Jura is the home of summer rain and mist, but this time the sun shone all day long every day for a week.

The hills were carpeted with heather and bracken with an under-lay of peat and you could hardly walk a kilometre along a track without seeing a hare, a hundred liquorice slugs or almost treading on an adder. Jura is a corruption of the Norse for Deer Island. They had the run of the place and the local landowners had been criticised for over a century for caring more for their stags than their tenants, but then, they would argue, the deer were far more lucrative. The human population had plummeted from 10,000 in the eighteenth century to only three hundred, while the deer had burgeoned.

The island is edged with long silver beaches on which the only footmarks were mine. My other life seemed elsewhere and out of reach. Communications were slow in coming. The telephone didn't arrive until 1949. It was a single line to Craighouse and when it rang you counted the rings and if they stopped at your number, you picked up the receiver.

A local told me that Craighouse had no street lights because 'they would ruin our view of the stars'. It was just a tiny cluster of cottages, the hotel and an enormous shed sporting an ugliness reserved for factories. There are many such 'factories' in the Hebrides; they manufacture whisky. Surprisingly, Scotch whisky is largely an Irish invention. The Distilling Act of 1779 taxed the output of Irish distilleries. Some firms went out of business because of the owner's sentimental attachment to his product, but the new law finished off many more.

Illicit stills soon produced two million gallons of illegal hooch over a single year and excise officers became the most hated men in Ireland. The Inspector General of Excise for Ireland was Aeneus Coffey. It occurred to him that there was a more efficient way than the old pot still that produced the spirit in small batches and at such low concentration it had to be redistilled. So he invented a still that delivered a continuous flow of high-strength spirit at the rate of 2,000 gallons a day. The spirit was so pure it had little flavour so the Irish distillers weren't interested. The Scots added a dash of malt whisky for taste and Scotch grain whisky came to dominate the market.

The Jura Distillery, of course, produced only dark malt whisky by some mysterious alchemy. The flavour comes from an unknown number of the five hundred different molecules produced by the fermentation and maturation of the brew. Even single malts are blended and their distinctive taste is produced by blenders who can 'nose' 2,000 samples a week. On the wall beyond the ranks of huge copper stills was a clock chart that codified what the blender was seeking or avoiding. The possibilities embraced 'peaty', 'smoky', 'rubbery' and 'rancid'.

Jura is mostly high ground, so you are always aware of the sea on all sides. Someone must surely have written on the psychology of islanders: why some feel imprisoned by the ocean while others are liberated. I had a wonderful feeling of being far away from the world, like a lone yachtsman, but without having to worry about trimming the sails or battening hatches. The clock ticked more slowly here and there would be time to concentrate on things – perhaps to paint or write a book. I wasn't the first to think like this.

The last eleven kilometres to Barnhill, George Orwell's old house on Jura, is a rutted track through wild country with the wind in its hair. Here he would indeed be 'extremely ungetatable'. In the bracken at the side of the track, we passed the rusty remains of Orwell's car where he had crashed it thirty-five years before.

The house was derelict when Orwell first saw it, but it became his retreat from the outside world. If he were to write a new book he must get away from the constant distractions of London. But why, I wondered, did he have to get so far away, hiding out here on the most isolated corner of an isolated island? There was no telephone within twenty-four kilometres and no shop within forty. Orwell had chronic lung problems and the island's only doctor was ancient and the nearest hospital was in Glasgow, twelve hours away, assuming the ferry was running.

But years earlier, during the war when life was precarious, Orwell had dreamed of a Hebridean island that he assumed he might never see, and good sense is no antidote for a dream. He also had a morbid fear of his adopted young son Richard growing up in the smoky city amid the coughing crowds.

He spent his time feeding the pig, making jam and chasing chickens. When not enjoying the fresh air, Orwell smoked incessant roll-ups of foul-smelling Nosegay Black Shag. His flock of sheep grew to fifty so he enlisted Bill Dunn to help him out. Bill had a wooden leg and nailed a cross piece to the 'foot' to prevent him sinking into the bogs,

The locals called Orwell by his real name, 'Mr Blair', and understood why anybody who wrote novels would wish to do so under a false name. They took to this notorious misfit perhaps because, like them, he kept to himself. Strangers were considered 'cannibals', but invited visitors from London were not uncommon. He never spoke to them about the book he was working on, although they heard him clattering away on his typewriter through the night, sitting up in bed in his seedy dressing gown. He considered writing a book was like suffering from a prolonged illness. Not something one would do unless driven by a demon.

The book was *Nineteen Eighty-Four*. It seems extraordinary that such a story should have been written in this idyllic place How could he create a world in which everyone is ever-observed by Big Brother, while he lived without neighbours, his activities noted only by herring gulls and ravens? Yet he almost called the book *Nineteen Forty-Eight*, the year in which it was finished. Later he claimed that, because of his illness, it was more pessimistic than he intended. It was also more prescient than he could have imagined. There are still dictators who warm to the notion of a boot stamping on a human face. And just as he predicted, we now have a national lottery, the breakdown of family and destruction of the forests, security cameras watching our every move and a form of newspeak so finely spun as to make it almost impossible for politicians to tell the truth.

A field beside the house ran down to the sea where Orwell

moored his small boat. He often went out fishing and tending his lobster pots. On calm days he would visit the deserted bays of white sand on the west coast and watch the seals. One day in 1947 he took two guests and Richard out on a camping trip to collect puffin eggs and brought them back in the boat through the Gulf of Corryvreckan that separates Jura from the island of Scarba to the north.

Here Gulf literally means 'that which engulfs'. It sometimes belched up a drowned boar or deer. A dummy weighted to have the same density as a human body and set adrift was immediately swallowed and when washed up the depth gauge on its wrist showed it had been sucked to the bottom one hundred metres down.

The whirlpool of Corryvreckan is classified as 'unnavigable' by the navy and should only be risked at slack water. Orwell's helper, Bill Dunn, is said to have swum across. He unstrapped his wooden leg, shunning the buoyancy it provided, coated himself in mutton fat and dived in. That was, of course, at slack water. Unfortunately, on this day Orwell had miscalculated the tide.

As soon as they rounded the point he knew they were in trouble. The surface was in chaos and immediately the boat was out of control. The outboard engine was snatched away and sank. By good luck they were swirled towards the tiny island of Eilean Mor. With the boat rising and falling four metres, Orwell's cousin leapt ashore clutching the painter, but the boat capsized. Three-year-old Richard was trapped beneath the hull until Orwell ducked under and hauled him out. Miraculously, they all managed to scramble ashore, having lost everything, even their shoes.

But they were still in jeopardy. They were soaking and stranded with the sea boiling all around them. They had no food or shelter and were lucky to be picked up by local fishermen only three hours later. They then had to walk home. When they arrived they were greeted with, 'What took you so long?'

*

I walked from Barnhill to the Corryvreckan. I could hear it roar long before it came into view. In a gale it is said to be audible twenty-seven kilometres away. No wonder locals say that 'Jura is a fine place to drown yourself'. The tide funnels into the strait and accelerates through at a speed of over fourteen kilometres per hour. There is a sill on the bottom that thrusts the water upwards to pile up on the surface. As the sill is shallower on one flank, the flow is deflected to form a vortex, the huge whirlpool on the Scarba side with a spectacular back-wash on the Jura shore. But worst of all, the bottom is so irregular that the racing current is flung every which way to form a chaos of terrifying swirls, standing waves, overfalls and deep troughs. This is water gone mad – one of the most dangerous straits in the world.

Surely it can't compare with the great whirlpool of Charybdis, the gulf that Virgil claimed 'swallows up the vast billows of the abyss . . . and lashes the stars with the waves'. Odysseus famously risked being engulfed in the days when Charybdis was difficult to avoid because, in attempting to give it a wide berth, you came within reach of Scylla who inhabited an adjacent rock. She had twelve feet and six long necks each, according to Homer, with 'a hideous head armed with a triple array of teeth set thickly and teeming with black venomed death'. Such was his fear of Charybdis 'that sucks down the tide thrice in a day', that Odysseus considered Scylla the safer bet. 'It is better by far to lose but six of thy comrades [which he promptly did] than all at a swoop,' he was advised. It was taken for granted that if Charybdis coiled around the keel of a careless vessel it would be drawn in and even Poseidon himself could not save it. Homer describes the sound of the whirlpool as being like a 'cauldron over a fire'. Curiously, the Gaelic name Corryvreckan means 'cauldron of mottled water'.

The evil reputation of Charybdis in the Straits of Messina was enhanced when occasionally mini-monsters were wrenched from the depths and dragged to the surface: bug-eyed hatchet fish and sabre-toothed viper fish, both with a luminous gape.

The straits where Charybdis swirls are, like the Gulf of Corryvreckan, a constriction that accelerates the flow of the tide. When the race of the biggest spring tides is opposed by a fierce wind, it can create a very choppy sea, but nothing to trouble a modern ship. Ancient vessels, however, were mostly small, open boats at the mercy of the wind or the skill of the oarsmen, so they were liable to be swamped, but unless Charybdis has changed its character in recent times, there was never a risk of being *sucked* to the bottom. It is of course possible that in such an active seismic region the bottom topography may have been rearranged so that the water is no longer deflected to form a significant whirlpool. Certainly Charybdis no longer turns within reach of Scylla's rock but about a thousand metres distant.

It was long believed that whirlpools gyrated above holes in the bottom of the sea. Charybdis was the sink hole through which the Mediterranean drained, for otherwise the adjacent coasts would surely have been flooded. These drainings, now miraculously desalted, re-emerged as springs and fountains on land. This fable was superseded by the idea that it emptied into the centre of the Earth and exited again in the Gulf of Bothnia. The word 'charybdis' was still used a century and a half ago to designate any imagined drain on the sea floor. As recently as the 1950s, the *Sailing Directions for the Coast of Norway* described the Maelström as being like so many other 'pits in the sea'.

The Maelström, off the Lofoten Islands in Norway, was even more feared than Charybdis. It was the epitome of the whirlpool at its most terrifying, at least in literature. Jules Verne made the horror clear: 'The Maelström! . . . a whirlpool from which no vessel ever escapes ... whose power of attraction extends to a distance of twelve miles.' Edgar Allen Poe's *'Descent into the Maelström'* was the final confirmation of the public's horror of an engulfing vortex:

> The mouth of the terrific funnel, whose interior, as far as the eye could follow it, was a smooth, shining and jet-black wall of water ... speeding dizzily round and round ... and sending forth to the winds an appalling voice, half shriek, half roar, such as not even the mighty cataract of Niagara ever lifts up in its agony ... The largest ships of the line in existence, coming within the influence of that deadly attraction, could resist it as little as a feather the hurricane, and must disappear bodily and at once.

Even the *Encyclopaedia Britannica* of the day confirmed the Maelström's 'prodigious suction', and the *Sailing Directions* warned that 'boats and men have been *drawn down* by these vortices, and much loss of life has resulted'. But like Charybdis and the Corryvreckan, the Maelström merely inhabits a tidal bottleneck and both the swirling flow and surface violence result from the irregularity of the sea floor below and the wind above. Even in the worst winter gales the flow does not exceed eleven kilometres per hour, which creates turbulence for sure, but not a gigantic all-consuming vortex. Local boats sensibly avoid the worst conditions, but fish the narrows while their boats waltz within the swirls.

It seems that fewer mariners have been sucked to their death than was once believed. Perhaps they appeased the whirlpools, for fishermen crossing the Maelström were said to throw in an oar to ensure they got over safely. This echoes the notion that Charybdis too could be calmed by flinging a log or a bundle of straw into the vortex, which could only swallow one thing at a time.

Being experimentally minded, I threw a Poohstick into the Corryvreckan and watched it rush away into the turbulence. It had little calming influence.

A Walk in
the Park

Carsaig Island, Sound of Jura

Argyll was once covered by the great Caledonian Forest. The Scot's pines didn't huddle together to form a dark wood, but seem to have spaced themselves out, as if planted in a park. Then 3,000 years ago the bones of the land were exposed when trees began to be felled for fuel or burned down to make pasture for sheep and to remove the hideaways for brigands and wolves. The last wolf in Scotland was killed in 1743. The purpled heather moors may be the glory of Scotland, but they also mark the tragedy of her lost forests.

Some great forests remain, but they grow beneath the waves. On shallow submerged rock off almost every shore are the jungles of the sea. Kelp plants may stand two metres tall with fronds as big as a tattered brown tablecloth. Another species of kelp may cling to their trunks and reach almost half a metre above the top of the main canopy. The forests sieve the waves and their huge surface invites colonists to settle. A hundred times more animals can live in the labyrinth of their root-like holdfast than on the adjacent rock.

I explored a wonderful kelp forest at Carsaig Island facing the Isle of Jura. In those days when divers were scarce, it was likely that I was the very first person to see many underwater landscapes. But not here at Carsaig. At this very spot, over thirty-five years earlier,

Jack Kitching became the first biologist to see the kelp forest at first hand.

It began in 1931 — in a dairy.

'I intend to descend to the bottom of the sea,' young Jack explained. 'So I will need some diving gear.'

'That would be advisable,' replied the sympathetic dairyman, wondering what this had to do with cows.

'Have you got a spare milk churn?' Jack continued. 'One large enough to go over my head?'

Jack bought the can, fitted a window, flanges to take the weights and a pipe to attach a garden hose. Wearing a rugby shirt and a pair of plimsolls, he strode into the sea off Carsaig. He trudged into the forest cutting his way through with hedging shears like a ghostly gardener. It was hard going at first as the plants were densely packed, but deeper down the forest opened out and he could walk between the kelps. At a depth of twelve metres 'it was possible through the misty water to see the "park" extending downwards much further on the steeply sloping bottom'. Kitching had taken his first steps to becoming one of Britain's greatest marine biologists.

These first dives were witnessed by a young student called Russell Lumsden, who had been invited to spend the summer with a friend's family at their bungalow on Carsaig. One day he wandered down to the shore and found two men 'frantically servicing two ordinary motor tyre foot pumps'. A few minutes later a man emerged from the waves wearing on his head an inverted 'ash bin' held down by huge lead weights front and back. They had to be removed immediately he surfaced for 'they were too heavy to stand up with in air. The safety theory was that if anything untoward happened in the deep, you just pushed off the ash bin and swam up to the surface!' That first day Russell helped with the pumping and within a week he was underwater wearing the 'ash bin'. He recalled 'relaxing on a comfortable rock and admiring the scene – the huge fronds of the weed with top shells grazing on them, soaring above me in the powder blue water. However, if one was down for more than twenty minutes or so, you were seized with uncontrollable shivering as soon as you came ashore.'

By this time the pumpers on the surface were already exhausted and they found they could often hasten the end of a dive if they made the diver suffer a bit by 'easing back on the air'.

Moored nearby was the *Coia*, an eight-metre yacht on which Jack slept. In those more proper days his fiancée slept ashore and every morning came to the pier, gave a blast from a foghorn and repeatedly yelled 'Coia – O! Coia – O!' much to the amusement of the locals.

Kitching had no time to relax underwater admiring the waving fronds; he was there to describe the forest and its inhabitants and to assess how they were affected by the waves, slope and silt. As kelps are plants he also measured the amount of light reaching them at different depths. His waterproofed photocell had interchangeable filters so that he could measure different wavelengths separately to see if the plants responded to the 'powder blue' light in the shallows or the green gloom that predominated deeper down.

He found that the canopy of kelp fronds cut out 80 to 99 per cent of the available light, leaving dense shade beneath. No wonder many beautiful light-hungry seaweeds perched on the tops of the kelp stalks to grab some illumination. Deeper down the available light changed tint as the water filtered out all red tones, and became progressively dimmer, but as the forest opened out into a park it cast less shade so the light levels under the canopy stayed remarkably similar at all depths.

It might seem that only the topmost layers of the canopy are in full underwater sunshine and any plants growing on the rock beneath are condemned to live in perpetual twilight. Not so, for kelps flap like washing in the wind. If you think a tree in a typhoon is having a hard time, you should see seaweed at the mercy of waves. But this motion is vital for the flora below. It parts the canopy and lets in light in brief flashes, and these momentary sun flecks sustain the plants below.

It is the same in terrestrial forests. Although, like seaweeds, most plants of the forest floor are shade-tolerant, the bulk of their photosynthesis comes from brief but frequent sun flecks. The more the canopy is stirred by the wind, the more light leaks down to the underflora.

Surely it is difficult to make metabolic ends meet if the light that sustains your growth and survival is fluctuating wildly all the time. No sooner does the metabolism get into gear than someone switches off the power. But that's not quite the way it works. Both the seaweeds and the forest dwellers are adept at using the brief flashes of light very efficiently – they grab all they can in the few seconds it's on offer. The point of photosynthesis is to harness light energy for processing carbon dioxide and most of this is 'fixed' *after* the flash of light has passed. Clearly, some vital aspects of photosynthesis keep running for a while once set in motion, so if another sun fleck follows soon afterwards, as it invariably does in the turbulent sea, photosynthesis may progress uninterrupted.

I sat out of the waves in the gloom under the canopy of kelp and watched the flickers of light that dappled the forest floor. Each time a sunbeam flashed, the red fronds of the seaweed garden seemed to burst into flames.

In Jack Kitching's day much of the equipment made the diver look like a gasworks on legs, but he had shunned such complicated ways to drown in favour of a milk churn. It was impossible not to warm to him.

My buddies and I went three times deeper than Jack had reached and passed through the zones of plants that he had been the first to describe from direct experience, but whose existence Edward Forbes had predicted in 1859. The kelp forest with its carmine companions reached down to almost thirty metres and then yielded to what Forbes had called the coralline zone, 'wherein the horny plant like polypidoms of hydroid zoophytes delight to rear their graceful feathery branches whose flowers are animals'. This particular 'polypidom' was a lumpy carpet of multicoloured anemones and dead man's fingers.

By now Kitching had become a world authority on the physiology of the tiny protozoa (microscopic animalcules found in water), and was using them as guinea pigs in his studies of how the working of cells was affected by high pressures. He had a steel chamber, appropriately called a pressure bomb, to produce pressures over a thousand times greater than that of the atmosphere. Once, when it exploded, a large fragment passed between Jack and a technician, missing them by centimetres before exiting through a wall.

Marine biology was his passion. Encased in a tweed jacket all winter long, he chafed to be free. As soon as the summer vacation arrived, Kitching departed for the shore. In rural County Cork he threw off his tie, donned faded shorts and the remains of a sweater savaged by sharks. At the first promise of sun, or even warm rain, he discarded his shirt to augment his freckles.

He spent almost every summer of his adult life at his modest, privately owned marine laboratory where I too would spend fourteen of the most wonderful summers of mine.

A Little
Night Music

Lough Ine, Ireland

When Frank O'Connor called Cork a hell-hole, he was corrected by a local who admitted that a hole it might possibly be, but hell-hole made it out to be far more exciting than it was. Honor Tracy thought 'the only truly impressive thing in Cork is its lunatic asylum . . . remarkable even for Ireland where only the pious and the unhinged assemble in any numbers'. Sean O'Faolain, a Corkman, was more cautious: 'It seems to be asleep. I have always found it to be asleep, but since it is part of Cork, it is safer not to believe for a moment that it *is* asleep.'

Deep into West Cork sits Skibbereen, which had a rather grander impression of itself. The *Skibbereen Eagle* once warned that it was 'keeping its eye on Russia'. On 2 September 1939, the *Southern Star* was unable to decide which was the lead story of the day and divided the front page equally between a breach-of-promise case:

ALLEGED TALK OF A 'MATCH'
Story of trips to Ballyvarney

and:

GERMANY INVADES POLAND
Hostilities along entire frontier, France mobilises

The unfortunate positioning of the photograph of Hermann Goering seemed to imply that he was involved in both escapades.

Just beyond Skibbereen is Lough Ine, one of the most beautiful lakes in Ireland, yet not a lake at all. From the shore you may notice that someone has hung the remnants of white handkerchiefs from the branches of the trees bending low over the water. Look closer. They are the bleached ghosts of sea lettuce. Kick off your shoes and venture into the shallows, but be careful where you tread for there are no reeds and rushes here. Instead, there are spiky purple sea urchins and aggressive devil crabs, arms outstretched, bursting for a fight. Don't be fooled by its kingfishers and otters -Lough Ine is an enclosed fragment of the sea masquerading as a lake. If you see the lough for the first time and are not entranced, feel your pulse, for you might just be dead.

In 1938 Jack Kitching and another young biologist, John Ebling, came to study the lough. They *were* entranced and spent the next forty years researching the ecology of the lough, making it the most intensively studied patch of sea water in the world. Their work at Lough Ine was to revolutionise marine ecology.

Apart from their eccentricity, they could not have been less alike. John was ebullient and bawdy. He had hordes of funny stories and told them repeatedly, with gusto. Jack was more reserved and a little puritanical, with no gift for small talk. He once sent a postcard from Lough Ine to his favourite research student. On the front was a colourful photograph of the lough, but on the back his entire greeting to her read: The hypolimnion is devoid of oxygen. Best wishes, Jack.

Jack's drive and John's organisational skills were the ideal combination, but in all other respects they were incompatible. The funny yet tragic story of their friendship and collaboration I have told elsewhere in *Reflections on a Summer Sea*. In it I also tried to capture the myths and the magic of the place and its stunning scenery, both on land and underwater.

The locals were warm and witty. When touring West Cork in a diesel car, a friend was relieved when the third garage he tried served diesel. 'It's very cheap,' he observed. 'Ah sure,' said the proprietor, 'I have a monopoly. I can charge what I like.'

Tourists came to admire the lough's ability to shimmer in the sunlight or turn moody and mysterious in the soft rain. But there are scenes that few visitors get to see. From the road you can't spy the secret exit to the sea – a channel so narrow that the tide rushes in and out like a cataract. The water takes on a steep slope and deserves its name – the Rapids. Four times a day the torrent slackens as the water level inside the lough matches that of the Atlantic Ocean outside. Although the flow reverses immediately, for fifteen minutes or so you can dive there. We studied the opening and closing of neon-decorated jewel anemones. This involved me diving into the Rapids to observe them both during the day and at night. They were the first night dives I had ever done.

As the current waned, I slipped from the warm night into the chill water and the embrace of the kelp. Beneath the canopy all was dark, but I could sense the movement of the seaweed straps above me. My torch revealed that I was caged by the vertical stems of kelp. Between them in caverns of darkness, the beam caught pairs of unblinking eyes all looking my way. I felt like a child trapped in a cartoon night jungle. All the eyes were disembodied. They hung there with nothing attached. Just eyes, no bodies. But something was moving. Closing in I saw it was the mouthparts of a transparent shrimp. It turned sideways and even its eyes vanished. But now I could see food passing through its gut. The only part of the creature that was visible wasn't part of the creature.

There were dozens of them as well as worms and tiny crustaceans that had emerged from between the kelp 'roots' in search of something dead for dinner. This was the night shift of animals rarely seen out during the day. For most of us, once we have outgrown our teenage years, the night is a quiet room into which we retreat to await the day. But the night is alive with insomniacs who doze when the sun is up. The prawns, for example, are nocturnal; this is their busy time. They would be far more conspicuous during the day when their enemies come out to prey. In contrast, the lough's purple sea urchins are the most conspicuous inhabitants of the shallows in daytime, and are so visible they are sometimes taken by gulls. But as the sun fades so do urchins. They hunker down between the boulders for safety – it is predatory crabs that rule the night. After a series of good years for the crab population, the numbers of urchins plummeted.

Starfish are also nocturnal feeders and deadly enemies of urchins and clams. If just the tip of a starfish arm touches a scallop, it jets away to safety. Even though crabs have good eyesight, like the starfish they stalk their victims by scent and touch.

I turned off the torch and, as my eyes accommodated to the dark, I realised it wasn't dark at all. A swimming worm left an iridescent wake and the kelp was outlined in a faint green luminescence. When I stretched my hand up into the flow, it too was etched by a constellation of cold stars. There was something in the water. Specks collided with my face mask, like blown sparks, and were quenched in an instant. These tiny silent fireworks of the sea were single-celled algae called *Noctiluca* (night light) and millions of them were drifting around in the water. On dark overcast nights it was not uncommon for the surface of the lough to glow with every smear of an oar. When I dived at midnight in a cave not far from the Rapids my entire body became a green spectre.

Lots of bacteria, algae and a variety of animals can produce light in all the colours of the rainbow from violet to fiery red. Some manufacture their own while others culture bacteria to do it for them. We also inadvertently harness bacteria. Sweaty individuals are politely said to 'glow' and there are reports that some really do.

The champions are the sea gooseberries, some shrimps and a

floating sea squirt called *Pyrosoma* (firebody). One night in 1872 the research ship HMS *Challenger* en route to Brazil passed through an enormous shoal of *Pyrosoma* that glowed like white-hot iron. 'The wake of the ship was an avenue of intense brightness,' wrote Charles Wyville Thompson. 'It was easy to read the smallest print sitting at the after-port in my cabin; and the bows shed on either side rapidly widening wedges of radiance, so vivid as to throw the rigging into distinct light and shadows.' His assistant wrote with his finger 'on the surface of a giant *Pyrosoma* as it lay on the deck in a tub and my name came out in seconds in letters of fire'.

Luminescence signals to potential mates in the dark or glows on the tip of an anglerfish's rod to lure unwary guests for its dinner. Some swimming crustaceans fire off flashes if pursued, just as a warplane releases chaff to ward off missiles. Deep-ocean squid squirt luminous ink to distract a predator; brittle starfish often shed an arm when attacked and then the detached limb glows alluringly while the rest of the animal limps away into the shadows. Some just produce light as a by-product of their chemistry and glow brightly in the privacy of their burrows or exhibit it only on their underside where it will not attract the attention of predators. It is as yet unclear why the shark called megamouth has luminous lips or the dong had a luminous nose.

The current in the Rapids grew steadily until my dive mask began to tremble. Just as the kelp blades were passively pulled downstream, I let myself go with the flow and was swiftly dragged out beyond the canopy of kelp and down into the opalescent underwater landscape of a creek.

There was a wrasse lying on its side on the sand. Dead I thought, but it was breathing and a prod brought it reluctantly to life. During the day wrasse may be more sociable, but at night they sleep alone. Some tropical wrasse bury themselves in sand when night falls and in the aquarium, if you switch off the light, they put themselves to bed.

Aquarists claim that each has its favourite sleeping place and returns to it every night.

I lay back beside the dozing wrasse and gazed up at the undulating surface of the sea where the newly emerged moon continually separated and coalesced like a dancing amoeba. A moon jelly emitting a faint blue light drifted between me and the moonlight and was instantly etched in silver. Every detail became hyperrealistic, yet unreal. I became entranced by the waltzing moon and held my breath to listen to the sounds of the sea.

Divers are deafened by the rasp of their regulator and the flutter of escaping bubbles, but if they held their breath and listened, *really* listened, they would hear the heartbeats of the ocean. At night, when shapes are lost in shadow and colours hide in shades of grey, sounds come into their own. Even though my ears were not designed to hear underwater, from out of the silvered shadows came the long sigh of water flowing down the Rapids, the lisping waves easing round a reef and the soft rattle from the shore as the undertow retrieved a pebble it had donated to the strand only seconds before. And there were other anonymous creaks and clicks, as if I were wallowing in an old wooden ship.

In water, sound travels four times faster and much further than on land. Dolphins and killer whales use sonar to detect the direction and distance of prey, but the whale often shuts off its sonar to listen for the splashing and breathing of distant creatures. We have forgotten how to listen. In Malaysia, some fishermen hang overboard with their bodies and heads in the water to listen for fish. The son is taught by his father and at first can hear little, but within a year he can distinguish the sound 'like the wind' of one species of mackerel from the 'parched rice' noise of another, and of course neither of these could be confused with the scabbard fish, which has a 'voice like a crow'. You don't have to have a practised ear to distinguish the swimming stroke of Audley's lethrinid, which sound like a 'pile of coins falling over', from the thump of a wrasse, which brings to mind the sound of a 'frenetic squirrel in a wooden box'.

On a tropical reef I have distinctly heard a parrotfish crunching coral and a spiny lobster chirruping like a cricket and using the same mechanism to generate the sound. The sea is alive with chatter: beds of mussels click like castanets, crabs and prawns can grunt and snap, and fish have been heard to make sounds like drumming, knocking on wood, the drone of a World War II bomber, a bottle filling with water and a trumpet blast. The various species of croakers are great at drumming, humming, hissing, whistling, purring, popping, snorting, squawking, creaking and of course, croaking. The loudest species was seen chuntering to itself eighteen metres down and was distinctly heard by someone on the deck of a boat above.

Microphones dipped into tropical waters pick up lots of grunts and hiccups during the day, but when night falls massed fishy choruses raise the sound level by as much as thirty-five decibels – the difference between a quiet street and rush-hour traffic. Fish listeners believe that these choruses reverberate over the entire 2,000 kilometres of the Great Barrier Reef and chorusing keeps the loose shoals together.

This only makes sense if fish can distinguish different sounds. In the laboratory, researchers used food rewards to train fish to distinguish John Lee Hooker from Bach. Thereafter, even at first hearing, they could correctly categorise Muddy Waters and Beethoven.

Divers beneath ice, which traps in underwater sound, complain of the cacophony made by seals. The calls of dolphins and whales are so well known that people buy their CDs, and whale-watchers are writing a dictionary of whale sounds and their meaning. And they don't whisper. Some whales are so noisy they would be refused take-off at Heathrow.

Dolphins mimic the calls of their pod mates. If one produces a distinctive set of whistles it is instantly copied by another hundreds of metres away. Oceanographers surveying the sea floor using a device that emits a swooping whistling, report something was imitating it and answering its call. Whales of the same species have local dialects, although they observe the same linguistic rules. When a couple of humpbacks from Western Australia wandered across to join a school near the Great Barrier Reef, the host whales took up the newcomers' argot and within two years their local dialect was extinct. An extra-ordinary feature of whalespeak is that it adheres to the basics of language. Humpbacks change their mating songs from one season to the next, but retain the structural elements and principles of compo-sition, while allowing the content to evolve.

There is great concern that human activities are endangering marine mammals. Sea-floor explosions used for seismic surveys can interfere with the dolphins' ability to deal safely with the nitrogen that builds up in their bodies during dives. As a result, they are suffering from the bends, until now a complaint almost entirely confined to human divers. Worse still, the much higher-powered low-frequency sonars now used by warships are having serious effects. The US Navy has admitted that their sonar was 'the most plausible cause' of the trauma suffered by whales with bleeding ears that beached themselves and died in the Bahamas. For the time being its use has been suspended by order of the US courts.

Shipping is claimed to be making the Pacific ten times noisier than it was thirty years ago and this might confuse animals that communicate and navigate by means of sound. But the seas have always been far noisier that we thought. We simply weren't listening.

Oceanographers opened their ears thanks to a military relic. The Sound Surveillance System was an array of underwater microphones that tracked Soviet submarines during the Cold War. When the system was superseded, it was donated to the oceanographers. Deep down where the microphones are moored, there is a 'sound channel' where sound waves speed along without getting scattered either by the ocean's surface or the bottom. They carry the pulse of ships and the cries of whales, but at lower frequencies they carry far more than that. The oceanographers have given them names that fit the sounds they make, like 'Bloop', 'Boing' and 'Gregorian Chant', but are only guessing at what makes them. Some last continuously for days or even years; the rising, bubbling tones of 'Upsweep' were heard continually from 1991 until 1994 and were thought to have come from a place where sea water encountered a pool of molten lava in a deep chamber. The rush of 'Train' seems to be generated by currents sweeping round a seamount halfway between Chile and New Zealand. 'Slowdown', when speeded up, sounds just like the pages of a magazine being riffled. So it may be the result of another type of friction – the deep growl of ice rubbing against the land. 'Bloop' is thought to come from a marine creature, but it is far louder than anything even the most raucous whale can make, so it must be a giant . . . something.

Who was it that called this the silent world?

Playing the Field

Lough Ine, Ireland

To observe nature and ponder what is going on is natural history, a valuable fact-finding mission. Measurements of the factors thought to be influential (perhaps light or wave action) quantify the information and lead the scientist to test the validity of the conclusions that by now are inevitably forming in his mind. We call these tests experiments and they are at the very heart of research. They usually involve a purposeful alteration of one or more of the supposedly influential factors (while keeping the others steady) to force the system to reveal how it works.

Most scientists consider the laboratory to be the only fit place for experimentation. It allows them to control what's going on and they can always pop next door for a cup of tea while the test progresses. But in the nineteenth century agriculturists thought that it would make more sense to conduct their tests on a much larger scale in the fields. It began in 1834 when Mr Lawes, a manufacturer of chemicals, decided to compare the growth of turnips grown entirely with farmyard manure with those nurtured by chemical fertiliser. He carried out the trials on his estate at Rothamsted and converted a barn into a laboratory. Today, Rothamsted is Britain's foremost agricultural research station.

In 1903, a rabbitproof fence erected at Breckland in Suffolk showed that rabbit grazing encouraged the growth of grasses, which grow from the base, and eliminated broadleaved plants that grow from the vulnerable tips. This was a real experiment in which grazing pressure was manipulated at several sites to ensure the results were repeatable, not just a fluke. Fledgling terrestrial ecologists in Britain embraced the idea of field experiments, but they found little favour in Continental Europe or the United States for another fifty years.

Nor were they adopted by marine ecologists until Louis Renouf, the Professor of Zoology at University College Cork, came to Lough Ine. In 1932, together with Kenneth Rees, a botanist, he published a brief outline of the research in progress at the lough. They described an attempt to understand the distribution of intertidal seaweeds by transplanting them to different levels on the shore, and also placing slabs of bare rock at different depths to monitor the colonisation of animals and plants and assess how they competed with one another. These 'field' experiments arose from their growing conviction that in the distribution of marine organisms biological factors rather than physical and chemical ones 'play an important, and perhaps decisive part'.

This would have been a pioneering attempt to carry out experiments in the sea but, as with many of the projects that Renouf planned, the results were never published. It was left for Jack Kitching to blaze the trail.

In his early years at Lough Ine, Kitching was obsessed with the Rapids. In 1947, he measured the currents and began to catalogue the abundance and distribution of the animals and plants. The water accelerated to a torrent of almost three metres per second in the middle of the channel, but slowed considerably towards either end. Since different species occupied different regions of the Rapids it was easy to spot those that 'preferred' the current and those that shunned it. But all this told him was *where* they lived, not *why* they were there. And the reasons were as varied as the creatures themselves.

Snails were abundant on kelp where the current was slow, so Jack exposed them to higher flows when attached to kelps or slates suspended in the Rapids. He showed that they were simply unable to cling on sufficiently tenaciously in fast-flowing water. On the other hand, seemingly delicate but firmly attached hydroids were most abundant in the swiftest current. When transferred to a calm bay nearby they became clogged with silt and died within a few weeks, while those in the Rapids thrived. The hydroids used as the comparison were not simply left behind, they too were moved, but then returned to the Rapids. Thus Jack could be sure that any differences the transplanted animals exhibited were due to their changed circumstances (calm not current), not to the distress of being disturbed, which *both* sets of animals had suffered. The results indicated that the hydroid needs water motion simply to wash away silt.

Beautiful jewel anemones also inhabited areas of moderately fast flow and Kitching suspected that again this was to avoid silt. So he moved anemone-covered boulders on to a calm, muddy bottom and placed some on spiked frames ('Gandhi beds' – Jack loved to name things) either anemones up or anemones down but well clear of the mud. All those facing up eventually died out, but those facing down, which were free from falling silt and shaded, all survived. Silt was still the main suspect until he investigated the responses of submerged anemones in floating boxes. The anemones exposed to sunlight closed up during the day and expanded their tentacles only at night, whereas those shaded by lids stayed open and feeding all the time. Simply taking off the lid during the day caused them to close. They also closed up when I removed part of the algal canopy in the Rapids and let in the light. At night they opened to feed, but within a few weeks the fluttering fronds of newly grown seaweeds kept brushing against their tentacles and they retracted. So jewel anemones enjoy a brisk current that brings a continual supply of food, but live in shady places where they don't have to compete with seaweeds and can stay open all hours.

Mussels also occurred in the region of greatest current and therefore least sediment, but they were also found in the lough where there was little water flow and lots of silt. When Jack transplanted them to other sites that seemed suitable but lacked mussels, they were consumed by crabs. Experiments in which crabs were confronted by mussels in submerged cages showed they had a taste for them and could break open even the largest shells with a whack from their claws. Consequently, mussels were confined to places where they could avoid break-ins, places where crabs were scarce because of strong currents or waves, or so calm that crabs were hard pressed to get enough oxygen for their energetic lifestyle.

Kitching and his long-time collaborator fulfilled Renouf's prediction that Lough Ine was 'a natural laboratory tank and . . . the solution to many problems may be worked out along its shores'. In this ideal living laboratory Kitching and Ebling not only showed that experimentation in the sea was possible and desirable, they also developed the methods that could be used to manipulate living systems and goad them into revealing how they functioned. Other marine ecologists, especially in the United States, began to follow where they had led. One of the most talented was Paul Dayton who was a graduate student at the University of Washington in Seattle.

Dayton studied numerous sites, from the wave-worried offshore reefs of the Pacific coast to the sheltered channels of the San Juan Islands. He attempted to assess the effects that all the leading players on the shores had on each other. Learning from Kitching, he protected small areas with barriers made from cereal bowls and dog dishes with their bottoms cut out. These were screwed to the rock to provide circular 'islands' free from grazing limpets or to corral various numbers of limpets inside and provide controlled amounts of grazing. These and other simple devices enabled him to assess the ecological effects of limpets, snails and starfish both separately and with all three acting simultaneously. His first scientific article became a classic of experimental marine ecology.

Field experiments are essential to prevent researchers embracing explanations that seem seductively convincing. For example, on the legs of the dock at the Friday Harbor Laboratories on San Juan Island, mussels covered the upper reaches down to the top of a zone of jewel anemones. Removing the anemones should have allowed the mussels to colonise the newly available space, but without the anemones the mussels died out. They were eaten by starfish climbing up from the sea floor. Starfish couldn't crawl up over a vertical carpet of anemones, but once it was removed the mussels were vulnerable.

The tide pools on the open coast were full of large anemones that fed on mussels, as did the ochre starfish. It seems obvious that if starfish were added to a pool they would filch mussels and the anemones might starve. In fact, they got ten times more mussels than before because the starfish dined out on the open rock and dislodged lots of mussels. These washed into the pools to be consumed by the anemones.

Both anemones and the sunflower starfish ate pool-dwelling sea urchins, so surely they must compete for food. Not so. If a starfish entered a pool, the urchins abandoned their secure depressions in the rock and stampeded away. The waves then washed them into the waiting tentacles of the anemones, which feasted on twenty times more urchins than before.

The moral is that in nature apparent rivals might be unsuspected allies. More importantly, the most obvious explanation may be wrong – so put it to the test.

Starfish were one of the main providers of that scarce commodity space, for they removed both of the main monopolisers of the rock, mussels and barnacles. On some shores removing all the starfish allowed mussels to become so abundant that virtually all the original inhabitants were eliminated.

Clearly some species are more influential than others and the diversity of life on shores could depend primarily on the activities of just one or two species. If ecologists are to predict the future, it is vital that they identify these 'critical' species and the extent to which they might be affected by, say, oil spills or climate change.

However, mankind's disregard for living creatures unless they are cuddly and the accelerating pace of environmental degradation may not allow sufficient time to identify those species whose loss would have immense consequences. By the time we discover their importance it may be too late.

Paul Dayton never forgot the debt he owed to the pioneering work of Jack Kitching. Years later he wrote to tell me that he had been to Ireland and made 'a pilgrimage to Kitching's little lab. For an old atheist like me it was about as close to religion as I could get.' I was about to do the pilgrimage in reverse to visit Dayton's research sites thanks to an invitation to teach at the Friday Harbor Laboratories. Sadly, this meant that, as it turned out, I would never work at Lough Ine again. But I would not forget those wonderful summers I spent with Jack at the lough learning to be a marine ecologist.

Alien
Invasions

San Juan Island, Washington

There are more than two hundred named islands and many more anonymous islets in the maze of waterways that separate Washington from Canada. All were ice-sculpted over 11,000 years ago and are now topped with firs and cinnamon-barked arbutus that cover the scars. Racoons climb the trees, otters and yellow-striped garter snakes forage on the shore.

The discovery of new lands often lay at the end of a long voyage. Subsequent visits served for trade or conquest. There was a freedom of the seas; they could be dominated, but not owned. New land, however, was up for grabs. In the still wild places I often felt I was treading in the explorers' footsteps.

George Vancouver, whose name adorns an island and a city, had been a fifteen-year-old able seaman on the final voyage of Captain Cook's *Resolution*. He returned nine years later in 1792 as master of the *Discovery* seeking the North-West Passage. While surveying the coast he felt compelled to name every bay, mountain and island. Some because they reminded him of England (Dungeness). Others rewarded his officers – Puget Sound, Whidbey Island and Mount Baker – or honoured influential superiors, including Captain (and soon to be Rear Admiral) Peter Rainier (Mount Rainier), and people of importance such as Marchioness Townshend (now Port Townsend). The Spanish had been here before him and already named the San Juan Islands and the Juan de Fuca Strait.

Sailing over an oily calm sea with bald eagles circling the rigging cast a spell on the crew. A young midshipman wrote: 'It had more the aspect of enchantment than reality, with silent admiration each discerned the beauties of nature, and nought was heard on board but expressions of delight murmured from every tongue.' Vancouver was also entranced, although he did not see a wilderness but a garden:

> A landscape, almost as enchantingly beautiful as the most elegantly finished pleasure ground in Europe . . . an extensive lawn . . . diversified with an abundance of flowers . . . and shrubs of various sorts, that seemed as if they had been planted for the sole purpose of protecting from the N.W. winds this delightful meadow, over which were promiscuously scattered a few clumps of trees, that would have puzzled the most ingenious designer to have arranged more agreeably.

Despite the occasional fog in which the world lost shape and distance, the summer was warm and clear and dry. Though the air was placid the sea was not. Big swells are unusual, but there is a surprise around every headland. The labyrinth of channels between the islands act as flumes through which the tide must squeeze, sometimes accelerating to sixteen knots to do so. What seems from a distance to be a fair passage suddenly transforms as you approach into a boiling, swirling chaos that swallows trees. It is a sea in panic – and so is the mariner. Vancouver named one such place Deception Pass.

The sailors caught glimpses of the sea creatures that haunt these waters: the pink flashes of salmon and the scimitar fins of killer whales, so common here that one of the San Juan Islands is called Orcas. But the explorers rarely came ashore and never intentionally went beneath the waves. What sights they missed.

The usual suspects are present, but in almost every case they are vastly bigger than their relatives elsewhere. The pools are crammed with apple-green anemones forty centimetres across that can swallow a sea urchin whole, a frighteningly phallic-looking clam, locally called the gooeyduck, weighs in at seven kilos despite its lightweight shell; a barnacle sits like a giant inkwell thirteen centimetres tall, the twenty-armed sunflower starfish has a diameter of 130 centimetres.

This is also the home of the giant octopus. It has a span of three metres (one seven metres across has been reported) and has been known to emerge from the waves to snatch a seagull. Underwater, it can jet around, but at the sight of a crab out in the open, it expands the webbed bases of its tentacles, parachutes down and envelops it like Dracula with his cloak. If the prey is hidden beneath a rock, it insinuates a single tentacle that can touch, taste and smell. It entwines the creature and drags it out. Although it can disarticulate even the biggest crab, instead it calmly drills in through the victim's Achilles eye and sucks him dry. There are lurid divers' tales of close encounters with giant octopuses. Every one of the storytellers had narrow escapes, but when underwater I prefer to believe the professor who said 'a farmer is more likely to be attacked by a pumpkin than a swimmer by an octopus'.

I came across one just sitting on the bottom. It seemed as if it were waiting for someone. At my approach, two large horns reared up on its head. Visions of man-eating pumpkins reared in my imagination and suddenly I decided I was short of air. Imagine our state of mind if the giant octopus stalked the night on land instead of the ocean. The slavering creatures that lurk in the dark recesses of movie spaceships are for me nowhere near as weird and wonderful as this alien life form.

I had come to San Juan Island to study another alien invader. In the early 1960s a Japanese seaweed, *Sargassum muticum*, had accidentally been imported with oysters on to the northern coast of France, but no one noticed. By 1971 it had crossed the English Channel. Plants were spotted on the Isle of Wight and they were spreading. Half the Boy Scouts on the south coast were mobilised to uproot the invaders, but without success. Extravagant claims were made that the weed would clog our harbours, impede our ships and supplant marine communities. The effects of introduced species are usually overlooked completely until they are seen as a valuable source of research grants. Then, it seems, the entire British way of life is threatened.

Some introductions do indeed become invasions. In Scotland, the aggressive wild *Rhododendron ponticum* is supplanting all the native shrubs and shading out the herbs. The island of Colonsay hired full-time cutters to destroy every rhododendron they could find. Ironically, in Turkey, its original home, the ancient forests of *Rhododendron ponticum* have been almost wiped out and it has been designated a threatened species. Invaders often get out of hand. In 1884, delegates at a cotton exposition in New Orleans were each given a pretty South American water hyacinth to take home. No doubt they treasured them for a while then dumped them into the nearest watercourse. In 1894, it was introduced into a botanic gardens in Java where it soon outgrew its pond. The floating weed didn't tire of invading until it had impeded navigation on every lake and river in the southern United States and throughout the tropics.

On the other hand some invaders are rampant to begin with then decline dramatically. At one time, introduced Canadian pondweed choked every canal in England, but even while it was still invading new sites it was fading away in the places it had first colonised and was becoming scarce. Would *Sargassum* be a continuing pest or just a flash in the pan? San Juan Island might hold the answer.

Sargassum muticum first appeared in Washington in the early 1940s, again probably arriving with oysters imported from Japan. Since then it had spread along the Pacific seaboard and would soon reach Alaska in the north and Mexico in the south. But it had been residing here in the San Juans for about forty years so this was the perfect place to assess its long-term impact.

The plants were abundant on the shore but not to the exclusion of others. It seemed to have become a respectable member of the community. Had it lost the vigour it was still exhibiting at the advancing edges of its distribution? No it hadn't. At higher water temperatures it still grew while you were watching and reproduced with excessive exuberance. Indeed it was the perfect invader. Large branches broke off and drifted away, sometimes surviving for months and travelling tens of kilometres. They became fertile en route and then ejected progeny when they eventually hit land again. In Europe it was also spreading rapidly and its colonisation of a new region was invariably preceded by drift plants washing up on the shore. The pioneers then consolidated their position by dropping hundreds of thousands of tiny plantlets around each parent to saturate the shore with colonists.

Sargassum muticum not only grows much faster than its potential competitors, it is also perennial. Its branches die back each year after reproducing, giving a window of opportunity for smaller plants to grow beneath. It can't tolerate drying out so most plants grow in pools or underwater, leaving the bulk of the open rock to others.

Curiously, its behaviour back home in Japan gave no hint of aggressive tendencies. There it grows to less than a metre long, and is only found just below low-water mark, whereas in the United States it can reach ten metres tall and grow at depths of twenty metres. This is typical of many invasive species. *Caulerpa taxifolia* is a green seaweed that was accidentally flushed into the Mediterranean from two public aquariums. In their original home in the Caribbean the plants rarely exceeded fifteen centimetres tall and made small patches underwater, but in the Mediterranean they grew to sixty centimetres and formed extensive beds.

Whatever curbs these two species back home is clearly lacking when they go abroad. Both *Sargassum* and *Caulerpa* contain noxious chemicals to deter grazers. After aeons of evolution some of the herbivores back home may be able to neutralise them and tuck in, but that is unlikely for grazers that have only recently encountered them. Neither species seems to be anyone's favourite browse.

When *Caulerpa taxifolia* was first spotted by a French biologist just below the public aquarium of the Monaco Museum of Oceanography, the entire patch occupied only one square metre; but while the authorities dithered and the French and Monégasque governments argued about whose jurisdiction it was, the invader put out runners much as a strawberry does and eventually formed lush meadows stretching from Spain to Croatia.

Meanwhile, scientists in Monaco claimed that the spread of *Caulerpa*, with its scant fauna, and the consequent regression of the rich seagrass beds were both due to pollution, and that the *Caulerpa* had not come from the Monaco aquarium at all, but immigrated into the Mediterranean through the Suez Canal. Genetic 'fingerprinting' soon proved that the plants were an aquarium strain. Recently, *Caulerpa taxifolia* has turned up on the coasts of San Diego in California and New South Wales in Australia. Both seem to have originated from aquaria.

Accidental introductions come by all manner of transport and ships are major culprits. The cargo of a boat bound from Calcutta to Cuba was found to harbour forty-two species of live insects and spiders — it was a touring ecosystem. Indeed, *two* ecosystems, for nobody thought to examine the *outside* of the hull. It cannot be a coincidence that many of the organisms that foul ships' hulls are cosmopolitan; they have been carried everywhere by boat. Then there's the ballast. Ships are trimmed by taking on sea water, which is carried to the destination port and voided along with all the creatures it contains. No less than 367 species were found in the ballast water of a ship docking in Oregon. No wonder ports and harbours are often the focus for accidental introductions. San Francisco Bay has welcomed no fewer than 255 exotic invertebrates.

Imported aquatic livestock are also a rich source of unwanted passengers. When a biologist examined the drain water used to rinse three shipments of oysters and tropical fish imported into Hawaii, it contained thirty-eight species alien to the region. The untreated water was pouring directly into the sea. In Britain, all the main pests of oysters have been introduced with foreign oysters imported for cultivation. In the 1950s, a large green Japanese seaweed was brought to the east coast of the USA with imported shellfish. It fouled the shells of cultivated oysters, caused them to be washed ashore by waves and clogged up the sorting machinery. The locals called it 'Sputnik weed', for they were certain the Russians had dropped it from a satellite to ruin their harvest.

Many species are imported for commercial cultivation. In Hawaii three species of oyster are grown in the sea. None are native species; they came from the United States, Australia and Japan. Such introductions can be successful and cause little environmental change, but no one can be certain in advance that this will be so. All introductions are unpredictable and the cumulative effects of successive invasions of different species are unknowable. We are playing ecological roulette and the stakes are high.

*

Much as he loved the local scenery, George Vancouver felt it lacked one thing, the hand of man, preferably that of an Englishman: 'The serenity of the climate, the innumerable pleasing landscapes and the abundant fertility that unassisted nature puts forth, require only to be enriched by the industry of man . . . to render it the most lovely country that can be imagined.'

I spent three summers on San Juan Island over a period of eleven years and have seen it change. When I first visited Friday Harbor in 1973 it was a gentle, sleepy place. Old timers were fond of saying, 'Hereabouts, once you get to eighty you forget how to die.' It was a place you went to put water between you and the world, and so people came in ever increasing numbers. There are now three million visitors a year heading for the islands of Puget Sound and the San Juans.

The population of the island has more that quadrupled since 1960. There are far more bars and touristy shops now, and realtors trying to encourage visitors to stay and inflate the cost of housing above the means of local people. The seafront properties resemble log cabins on steroids. Each has a runabout moored to a private jetty. Out front a man was raising a flag on a pole. My two-year-old daughter spied him and shouted, 'Is *that* a redneck, Daddy?'

Inevitably, there was resentment against incomers from the existing residents, many of whom had several part-time jobs to make ends meet. When there is a shortage of money barter flourishes. The local notice board embraced a weird selection: '20 years of the *New Yorker* back numbers in exchange for solar panels'; 'Massage in return for goat husbandry'; 'Call Bill if you find a stranded whale and don't know what to do.'

Native Americans are rarely seen in town, though they have nine reservations in this region. The early explorers were unimpressed by the local natives. The nineteenth-century German naturalist Seemann was clear that 'man in a savage state exists in all his grossness . . .

uncivilized man, with all his ingenuity, cunning, and skill of his class, seems in general to revel in filth'.

Nor was he impressed with their artefacts. It is ironic that replicas of what he called their 'rudely made chests' and 'grotesquely carved figures and masks' now sell for several thousand dollars in the posh boutiques of Vancouver and Seattle.

We have no way of knowing what the natives made of the explorers. They must have seemed like visitors from another planet. Later, 'Anglos' washed up on these shores in waves, just as the 'native Americans' had migrated here from Asia at the end of the last ice age. The Indian tribes were allocated reservations and their rights were enshrined in a series of treaties, but these were resented by the white folk, as the historian Nordhoff makes clear: 'How many millions of feet of first-class lumber are sacrificed to provide an Indian rancher with huckleberries?'

Soon the terms of the treaties were conveniently forgotten and, in the waters of the San Juans, Indians were arrested for fishing 'illegally' where they were allowed to fish. Then in 1974 the Federal District Court decided to honour a treaty of 1855; the very document that had cleared them from the islands also gave local tribes the right to harvest fish and shellfish 'in common' with non-Indians. Judge Boldt interpreted this as meaning fifty-fifty, which brought a financial boon to the tribes, but fostered bitterness in the redundant white fishermen. There were interminable wrangles in court and even gunfire on the fishing grounds. There was talk of racial war.

Fuel was added to the flames when the Makah tribe resumed whale-hunting, which had once been a tradition. When it was rumoured that Native Americans would be given rights to the tidelands on the shore, it reddened necks even more. The tribes had at last received justice, but sadly at the cost of inflaming resentment. And they had been given greater access just as the fisheries were declining as a result of overfishing and pollution. Some salmon runs were now only a tenth of what they had been, and the tribes became rivals for the best grounds. The once rich shellfish beds in the shallows were no longer productive because of contaminated run-off from the land.

Long ago Chief Seattle of the local Suquamish tribe entreated the settlers to care for the environment as the Indians had done. He lamented the loss of his people: 'There was a time when our people covered the land . . . but that time long since passed away with the greatness of the tribes that are now but a mournful memory.'

He also issued a warning: 'At night when the streets of your cities and villages are silent and you think they are deserted, they will throng with the returning hosts that once filled them . . . The white man will never be alone.'

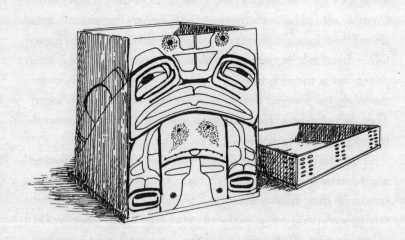

Pretending to
be a Leaf

Santa Barbara, California

A new administrative assistant at the Friday Harbor Laboratories was making an inventory of all large items of equipment. She was uncertain how to value a 'marine railroad'. In 1961, Mike Neushul decided to study the growth of seaweeds at different depths. A little promontory close to the labs sloped evenly down to twenty metres. Unfortunately, a swift current made it tricky to stay put in one place and the giant ferries that thundered past only a hundred metres away deafened the diver and turned the water to muesli. So Mike decided that instead of going down to the plants, he would bring the plants up to him. He laid railway tracks down the slope and attached the seaweeds to carts that were then lowered to the required depth. At intervals they were hauled up to be measured. Mike left for Santa Barbara two years later. By the time admin tried to value his railway, it was a liability not an asset.

Mike believed that every marine biologist should see beneath the waves, for how otherwise could they appreciate how marine creatures lived? As a student he swam competitively and played water polo for the university, so to get his protégés fit for diving he took them to the pool to swim laps and nobody left for lunch until the prescribed distance had been covered. It was as if they had enlisted in the Marine

Corps when they thought they had only signed up for the conservation corps.

Mike had planned to become an elementary school-teacher but, after a few sessions of 'pretending to be a leaf', he had second thoughts. Instead, he studied at the University of California at Los Angeles and a charismatic lecturer got him hooked on marine plants. In 1955 he began his PhD, but his studies were constantly interrupted so that he could go on diving expeditions.

Following an oil spill in northern Mexico, he rushed down to take part in the environmental assessment. On his last day he smashed his glasses. His eyesight was so bad that he couldn't see to tie his shoelaces. Fortunately, he had a spare pair; unfortunately, they were welded to the inside of his diving mask. So he wore the mask home. The arduous six-hour drive was not helped by having to continually wipe away the misty condensation. He made it but confessed, 'You should have seen the look on the faces of the US customs.'

In 1957 Yale Dawson, a leading taxonomist, enlisted Neushul to describe the flora of the kelp beds of southern California and northern Mexico. They made 240 dives together. The delights and distractions of trying to work underwater were just being discovered, as Dawson's account in the *Pacific Naturalist* reveals: 'The bewildering array of interesting and beautiful plants and animals, attracts the diver away at every hand from any attempts he may make at methodical investigation.'

Mike became the first person to dive in the Antarctic. In January 1958, he accompanied the Argentine Navy on an expedition to the South Shetland Islands, just offshore from mainland Antarctica. His aim was to determine if it was feasible to make repeated dives in the frigid Southern Ocean. Over ten weeks he made thirty-three dives, sometimes spending up to an hour underwater at depths down to twenty metres.

It was an eventful trip. En route from Patagonia in a small plane, he glanced out of the window to see the engine was on fire and they had to make an emergency landing. When eventually they arrived at the South Shetlands, strong winds and snow confined them to the huts for days, making diving 'a hideous prospect'. Even when it was fine, there were arduous treks across snowdrifts and fields of crystalline ice, carrying lead-weight belts and metal aqualung cylinders too cold to touch for fear of freezing to them. He did several dives a day, sometimes wearing his rubber suit for six hours at a stretch.

To dive under ice is to enter an alien world. From the brilliant glare of the surface snow you descend into the aquarium gloom below. Above you is a jade ceiling of solid ice or floating slabs called 'brash' or 'growlers' from their characteristic sound reminiscent of a giant grinding his teeth. The combination of crepuscular half-light and an underwater visibility of thirty or even an astounding three hundred metres, is unreal. The water column is decorated with streamers of penguin guano, and lazy icebergs grind by, gouging trenches in the sea floor metres wide and kilometres long.

The cold-hardened rubber seals on Mike's aqualung leaked and the valves threatened to freeze because as the compressed air expanded it cooled even further.

Mike tested the limited types of rubber suit available in those days to see which gave best protection from the cold. The water was $-1.5°$ centigrade; the temperature at which fish blood should freeze. This was somewhere far below mere cold, a place where the human body loses heat over twenty times faster than in air of the same temperature. Without insulation he would lose consciousness and be dead within minutes. The slightest leak at the neck or cuffs of his suit and he felt the excruciating stab of water; the tiniest gap between his dive mask and rubber hood conferred an immediate migraine.

Amazingly, seals retreat into this icy water to get out of the biting wind. Mike never thought he would envy an animal that is four hundred kilograms of blubber.

The big Weddell seals fight each other underwater and sometimes the posturing is abandoned and they go for the throat. In a black rubber suit Mike looked very like a rival seal. When diving from the shore, a surface lookout with a warning whistle kept an eye open for the three-metre long leopard seals, the most feared of all. Mike saw one take a penguin in a single gulp. Their reputation as man-eaters is undoubtedly overrated, but all seals can bite and even if it's a love nip it might still crush our puny bones. As it turned out, the most risky moment was trying to get out of the water through moving sheets of jagged ice when a hunter with a rifle took a shot at him thinking he was a seal.

He never forgot the easy grace of diving penguins and the rich underwater flora that thrived in the most inhospitable sea in the world, but it must have been a relief to return to the merely cool waters off California.

Mike was always anxious to acquire new techniques and as soon as he was awarded his PhD, was off to London on a fellowship to get to grips with electron microscopy. He also got to grips with a local girl, Susan Gardner, and they were married before he returned home.

He was appointed a professor at the University of California at Santa Barbara in 1963 where he indulged his fascination with technology, applying it to previously intractable problems in marine ecology. As a friend put it, 'When you get a new hammer, everything begins to look like a nail.'

Even on his earliest surveys he used a waterproof tape recorder to record his observations. When wearing a home-made microphone, he announced to the student with whom he was diving that he was about to give the first ever underwater lecture on ecology. The words were clear enough providing the student didn't breathe. As the lecture progressed the attention span of the breathless student became shorter and shorter. Mike rarely went on a dive without bringing a new device and if his students decided that the old way was better, his disappointment was palpable. His most extraordinary invention was an underwater microscope, but it worked.

He was certain that the commercial exploitation of useful chemicals from seaweeds was likely to increase. Indeed, he was to extract several antiviral agents from algae and also patented a method for treating HIV. To him, seaweeds would become a crop, so he was determined to improve the design of underwater farms. With a Japanese collaborator he also hybridised different species to produce faster-growing strains.

Mike had chosen to be an ecologist not a schoolteacher, but he became an inspiring teacher of students. He was bursting with ideas – some of them crazy, but all of them novel.

He devoted his energies to the study of the big seaweeds. They are extraordinary plants constructed very differently from those that adorn your garden. They have stems quite unlike the sticks on trees, their rootlike base is certainly not a root and as for their leafy frond, well . . . it's only pretending to be a leaf.

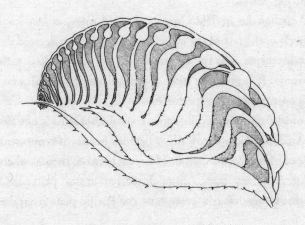

It's a Small
World

Isla Vista, California

Santa Barbara is a beautiful, sedate place where the retired go to be close to their parents. It is where Ronald Reagan retired to his Rancho del Cielo (Ranch in the Sky), awaiting departure to that final ranch in the sky. Oldies wear T-shirts that boast of 'Just another lousy day in Paradise'.

Downtown is a place of bright façades and shaded arcades inhabited by boutiques. It is a historic place and even the local McDonald's has a plaque on the wall boasting that it was here in January 1972, that the first ever Egg McMuffin was offered to an unsuspecting public.

An earthquake in 1925 spared the old mission, but knocked down much of the rest of the town, so the city fathers decreed that all new developments near the centre should be white-walled with red pantile roofs. It now looks like a beautiful version of Mexico, but smelling only of expensive perfume.

As most Americans would agree, coastal California is not like the rest of America, and geologically it belongs not to the continent but to the floor of the Pacific Ocean. The seam between them is where the San Andreas system slides. Although the two crustal plates are both edging slowly northwards, every time the Pacific plate overtakes on the hard shoulder there is an earthquake. Only a week before my family and I arrived, there had been a serious slip on the Santa Ynez fault that awakened Santa Barbara. The shock wrecked a couple of supermarkets, burst the fire hydrants on the street corners and shook up the locals. After-tremors were still frequent, but you soon got used to them. One morning, we were awakened by the apartment shaking and the bed swaying violently from side to side. 'Go back to sleep,' I said reassuringly. 'It's only an earthquake.'

We lived in Isla Vista student village, which was the antithesis of Santa Barbara. More shops sold incense and psychedelic posters than groceries. This was where the young could have a great time; it was a place for sex, sects and skateboards. Our neighbour, presumably a botanist, was growing a marijuana forest on his patio. Propped against the wall of the lecture hall were dozens of surfboards and as soon as class was over the students rushed down to beach to catch a few waves. It wasn't the best beach for surfing, but it was great for posing and getting tanned.

However, the anticipated golden beaches were marred by gobs of tar and one shore was half carpeted with the stuff. On the exit paths from the beach there were footbaths so that you could attempt to clean off the gunk. I assumed the tar came from the oil rigs that decorated the horizon. Then I went diving not far from the campus and was astonished to see that here and there the sand would suddenly lift to release little gobbets of black oil that rose to the surface. In one place there was an oozing mound of tar. This seep releases about sixty barrels of oil a day, and it has been estimated that a tenth of all the oil on the surface of the sea comes not from leaky tankers but from seeps such as these.

It had always been so. Long before the white man came, the local Chumash tribe carved soapstone figures, painted enigmatic pictures on the walls of caves and utilised what the ocean discarded. With axes made from the bones of dead whales they split planks from washed-up trees to make boats. They drilled the ends of the planks with bone augers so that they could be lashed together, and then filled the holes and cracks with the tar that the sea provided. They were the only tribe to make boats from wooden planks.

I also saw little bubbles of methane gas escaping from the sand. In the open ocean, massive escapes of methane from the sea floor have been accused of suddenly negating the buoyancy of ships passing overhead, by replacing dense water with gas, thus probably explaining sudden sinkings in the Bermuda Triangle and elsewhere.

Surprisingly, although oil is toxic, marine biologists found as many species in the seepage areas as elsewhere, perhaps because mats of bacteria utilising the fresh oil became food for animals that like that sort of thing. But chronic exposure to low levels of a substance is not the same as sudden and acute pollution. In 1969, one of the local oil rigs had a blowout and crude oil was carried ashore by storms. Damage to the kelp beds was surprisingly light, but hundreds of volunteers tried in vain to save the oiled cormorants, ducks and grebes by washing their feathers in detergent. The total amount of oil washed ashore may have been as little as 10,000 tonnes, only a tenth of that typically carried by oil tankers, but TV images of dead seabirds smothered by black tar had a big impact on the public.

Mike had gathered around him a cluster of talented young researchers and one senior citizen, who was not a typical pensioner. Every Wednesday I played tennis with Alex Charters, a world authority on ballistics. He had developed the ingenious instrumentation that monitored the aerodynamic performance of projectiles in flight. In 1944, he designed the first aeroballistic testing range, described as a 'national engineering landmark' by the American Society of Mechanical Engineers. The Secretary of War awarded him a 'Commendation of Exceptional Civilian Service' and after his retirement he was elevated to the 'Hall of Fame' of the US Army Ordnance Corps. There could not have been a more gentle man involved in the war effort.

Now with time on his hands, he eagerly accepted Mike Neushul's invitation to join the team. Alex knew little about marine biology, but understood the physics of flow, and that marine creatures all lived in the rush of tides and currents. It is moving water that brings nutrients to them, so Alex designed a watery 'wind' tunnel in which plumes of dye revealed the path of the flow around a kelp blade. They showed that the spines on the edge acted as spoilers to disrupt the flow and swirled the water into tumbling vortices that enhanced the plant's ability to absorb nutrients.

Although Mike's group studied giant kelp, one of the largest plants on Earth, this plant is only half the story. In season, it liberates millions of tiny spores and each germinates into *microscopic* male or female plantlets. If it were not for the tryst between their eggs and sperm, there would be no giant kelps. Every living creature starts life tiny and in the sea it is these microscopic stages that are the emissaries of sex and the agents of dispersal. These were the processes I had come here to investigate.

So I assembled a 'water broom' that Alex had designed. It sent pulsing waves of water across a submerged platform in a large aquarium tank. I injected fertilised eggs of *Sargassum muticum* into the flow so that they were flung across the platform, much as they would be swept across the sea floor. The platform was covered in surfaces with different degrees of roughness, like different grades of sandpaper.

The sticky eggs settled in the tiny concavities of the 'sandpaper' but only in those just a bit bigger than themselves. They could of course get into larger pockets, but were then plucked out again by the scour of the flow. Their fate was almost entirely a matter of hydrodynamics.

I thought that a film of slime like that covering most surfaces in the sea would prevent the eggs from sticking, but the opposite was true. The mucus mattress cushioned the landing so instead of rebounding the egg stayed and there was instant adhesion. I also tested the tenacity with which developing plantlets could adhere to the substratum by trying to blast them off with increasing velocities of water flow in the broom.

We have little feel for the world inhabited by microscopic organisms. Because of our size and power, moving through air is a doddle; it seems insubstantial. But put your hand out of the window of a speeding car and feel the pressure on your palm.

A sprinter streaks over the line and because of his momentum he takes many metres before coming to a halt. For minute larvae or seaweed spores, momentum has no meaning. The instant they stop swimming they stop moving. For them, water is not that soft runny stuff we know; they swim through a substance seventy times more viscous than air – imagine breaststroking through syrup or tar. Some pirouette so as to screw into the water, because it is so difficult simply to brush it aside.

Flowing water is also interesting stuff. It swerves around obstacles like boulders or seaweeds and in doing so accelerates and then breaks up into a wake of cascading turbulence. There is chaos in every wave, but at the rock surface surprising things happen. The lowermost molecules of water cling to the rock so tenaciously that they are absolutely stationary. The molecules above are slowed down by their attraction to the stationary water below. Adjacent water layers are also retarded, but progressively less so with increasing distance from the surface. The result is a layer of sluggish water beneath even the most raging flow. It is usually much less than a millimetre thick, but for tiny larvae or spores that have settled on the rock it provides a temporary haven 'out of the wind'. This gives them time to consolidate their anchorage before they begin to grow upward and into the full force of the flow.

Without an understanding of hydrodynamics we will never get to grips with many of the least known yet most important aspects of the life of marine creatures – their ability to disperse and to colonise new sites.

Alex Charters had a way of explaining the mathematics of flow that made me believe I understood it. He also had a six-metre motor boat and invited me on a trip to the Channel Islands. Before I could inform him that Jersey and Guernsey were just off the coast of France, he revealed that his islands were in the Santa Barbara Channel.

Only Skin
Deep

Santa Cruz Island, California

Forty-eight kilometres off shore, a necklace of large islands protects Santa Barbara from the Pacific swell. It took a while for the even the largest island, Santa Cruz, to rise from the mist.

Cabrillo, who claimed California for the Spanish in 1542, over-wintered his expedition on the islands, died from an infection and was buried there. His crew camped beside a Chumash village and found the inhabitants to be gentle but impoverished and living entirely on fish. By the 1880s the dwindling numbers of Chumash were devastated by frequent visits from Yankee whalers and Russian hunters. These vandals not only did for the whales and sea otters, they also killed the native men and raped the women.

Much of the island would be declared a wildlife reserve and become an everlasting fragment of how California looked before it was obscured by billboards and telegraph poles. There are several species found only on the islands: the dwarf fox, speckled skunk, and long ago, before it was hunted to extinction, the pygmy mammoth. Some evolved here in isolation, but most are the only survivors of creatures that once walked the mainland.

*

Ravens and bald eagles circled above the ninety-metre cliffs as our little boat approached. Every ledge seemed to be occupied by gulls and lower down within reach of the spray were oystercatchers and guillemots.

There was a great gash over forty metres tall in the cliff face. It was the entrance to the biggest sea cavern in California, the Painted Cave, so-called because multicoloured minerals have daubed its ceiling. I was eager to get in the water and didn't hear Alex's warning. I snorkelled into the entrance and round a bend in the cave, until all I could see above me was a dark vault. With a tranquil sea the immense cavern had the hushed reverence of a cathedral.

'Are you all right?' Alex shouted as I receded into the gloom.

Before I could turn to respond, the silence was shattered by barking and baying and then great splashes as unseen creatures dived into the water. The sudden cacophony in the echoing darkness was overwhelming. Shakespeare whispered in my ear, 'Be not afeared: the isle is full of noises.' What did he know about the anxieties of divers? Then they were all around me in the darkness – big black shapes in the water sliding past at speed. They were sea lions – slick-skinned laddos looking for trouble. The big males are territorial and I was a strange male intruding. They were trying to intimidate me and they were succeeding.

I finned gingerly towards the exit and was relieved to regain the safety of the boat. On another trip to the islands I would be circled by reef sharks, who have a reputation that any aspiring man-eater would envy, but would feel less vulnerable in their company than I did being buzzed by dozens of seals in the darkness of the Painted Cave. Nor would I return, for sea lions have the best memory of any creature other than humans and I was a marked man.

'Now you know why they're called sea lions,' Alex said calmly. 'They can certainly roar.'

It was well known, but not to me, that perhaps as many as a

hundred sea lions lived on a great shelf within the cave and at this time of year thousands more congregated on the islands to breed. As I could hold my breath underwater for only a minute or two, it was impossible not to be impressed by these sleek and absolutely aquatic air-breathers.

We cruised around the coast to where there was a wonderful forest of giant kelp and I snorkelled down to be part of it. Guillemots were arrowing into the sea from above. They didn't just stab the water then rise straight back to the surface, but swam around looking for fish. They have been seen casually swimming by at a depth of ninety metres. I dived down a modest seven or eight. A large ray flew between the stems of kelp with languid undulations of its 'wings'. Forget the plucky guillemots — fish are the true birds of the sea and nowhere was it more apparent than in this graceful gliding ray.

Interspersed between the kelps were the biggest plants of *Sargassum muticum* I had ever seen, at least eight metres tall. Every blade of seaweed looked newly minted. Nothing as soft and succulent as these plants could survive on land without a plague of locusts or caterpillars munching them to tatters.

Why are there no insects in the sea? Surely the most diverse and numerous creatures on the planet had missed an opportunity by ignoring the ocean. Some insects can swim, a few can dive, and others have snorkels through which they breathe. But what works fine when living in puddles and ponds could prove fatally inadequate in the turbulence of the ocean. Salt water also presents physiological challenges, but some insects thrive in salty inland lakes so that can't explain their sparsity in the sea.

Most marine animals breathe with the help of gills, feathery structures with an enormous surface area suspended in a stream of water often created by the animal's own motion. Absorbed oxygen is efficiently taken up into the blood that suffuses the gills. But insects have a very different system. Their crusty outer cuticle is perforated by tiny holes leading to narrow, much branched tubes that ramify throughout the body. The concertina action of the abdomen squeezes air in and out of the tubes aerating the tissue close to the finest tips of the tubes. This works for pumping thin air, but would be hopeless for dense water. One species of insect that lives in tide pools on the shore has evolved a sort of gill, so clearly it is not beyond their capabilities. So why didn't they go for gills and come to swarm in the sea as they do on land? Probably because they arrived on the scene too late. By the time insects burgeoned on land, crustaceans had already been exploiting every suitable niche in the sea for 100 million years, and crustaceans are equivalent to water-breathing insects. They include not just the familiar crabs and lobsters, but also the marine equivalents of grasshoppers, fleas and beetles. Anything that insects can do, crustaceans can do better underwater.

On the return journey the sea was unnaturally calm and I noticed some things skittering away from the prow of the boat. I asked Alex to cut the engine. There were clusters of insects on the silky smooth surface. When I disturbed them some hopped and others skipped rapidly away. I had seen bugs like this on garden ponds back home; these were pond skaters that had run away to sea. They were the ocean strider, *Halobates* (salt treader), which lives not *under* the sea but *on* it.

Water is a mass of molecules kept in liquid form by forces akin to magnetism. But at the surface where there is only insubstantial air above, these molecules cling to the adjacent water molecules even more tenaciously to give the sea an elastic skin. Only mercury has a stronger surface layer. It could be punctured by the finest needle but not by the water strider's needle feet, which are coated with water-repellent wax. They dance across this film like a ballerina tiptoeing on a plastic mattress. Four of their six legs are spread-eagled to distribute the weight and prevent themselves from being plucked up by the wind. I could clearly see the dimples in the surface skin of the sea beneath each foot. The long middle legs are used for rowing and the blades at the end of each are tufts of hairs, giving a whole new meaning to feathering oars. They create tiny waves that provide the purchase needed for the other legs to skate across the water. Every so often a line of the striders raced away at speed as if in rowing heats.

The greatest danger they face is getting dunked, for once the body is wet the bugs no longer float. When a swimmer emerges, a trivial film of water clings to his skin, but an insect would have to lift several times its body weight. As striders cannot swim or fly, to be on the safe side the body is held aloft and is waxed and silvered with water-repellent down that traps air. Even so, if I had to stay dry to stay alive, I wouldn't have chosen to stand here on top of the Pacific Ocean. The swell of the great ocean is no problem to them, just an enormous roller-coaster ride, but how, I wondered, did they survive a storm or a torrential downpour? Many don't, of course, and they are washed ashore in thousands after a gale.

The water strider is said to suck the juice out of jellyfish, probably not because they are really good to eat, there is just little else around. They in turn supply lunch for the occasional purple snail that drifts by attached to a balloon of bubbles.

Striders must have sex in full view for there is nowhere to hide, yet no one has seen them at it. The eggs are laid on flotsam such as wood and lumps of tar or, for the more discerning lady strider, floating feathers.

The ocean is a place of paradoxes – a two-dimensional world that covers seven-tenths of the entire surface of our planet, yet is almost unknown. It is by far the biggest sparsely inhabited desert on Earth, yet is carpeted with the world's largest reservoir of water.

The water strider has it to itself. Well, almost, for there are a few other surface drifters, and no sooner had we restarted the engine that we came across *Velella* (little sail), the beautiful by-the-wind sailor. It's a sort of jellyfish, an oval raft the size of a tablespoon and fringed with tiny tentacles. On top is a little triangular sail to catch the breeze. The sail is angled and all these were left-sailing; elsewhere in the Pacific they slant the other way and are right-sailing so that they follow the circulating currents of the ocean.

At first we saw only one or two *Vellela*, but soon we were passing through fleets of them, thousands upon thousands. For as far as I could see, there were tiny blue sails rising and falling in the swell and hoping for a fair wind.

Here Be Giants

La Jolla, California

The best way to be overwhelmed underwater is to sink lazily into the tall forests of giant kelp. The plants can be sixty metres high and grow faster than any other plant on Earth. Their huge canopy of translucent amber 'leaves' covers the water's surface, casting the world beneath into a blue-green gloaming. This forest is every bit as imposing as any jungle on land and far easier to penetrate. You can glide beneath the luxuriance of fronds, slide between the stems and hide among the shadows below.

Giant kelp is a superseaweed. Each plant has many hundreds of leaflike blades to capture sunlight. They expand at a hectic sixty centimetres a day and then, after five or six months, turn yellow and are shed like autumn leaves, only to be replaced by dozens of new blades eager to receive the sun. So the forest waxes and wanes with the seasons and regenerates itself twice a year. But in the shadows below there is more permanence, for each plant has a perennial anchor appropriately called a holdfast. It is a cone of tangled rootlike fingers and I saw one holdfast that was a metre and a half tall and five metres wide.

The comparison with a jungle doesn't begin to do justice to the kelp forest. With giant trees, the canopy of leaves perched almost out of sight above may be trembling in the breeze, but below are only great wooden trunks that hold the entire heavy structure upright. They remind me of the ponderous, fat-bottomed columns that kept up the roofs of ancient Egyptian temples, the best that could be done before the invention of soaring arches. In contrast, the seaweed forest is held aloft by nothing but air, because the massive plants are not supported from below but suspended from above. Each blade has its own small gas-filled float and the combined buoyancy of innumerable little balloons is immense. I dived beneath a boulder well over a metre in diameter, which was held clear of the sea floor by the kelp attached to it.

The stems of the giant kelp are pliant, more like lianas than trunks. So the entire plant yields to the current, with its thousand fronds unfurling and fluttering like pennants in a gentle wind. This forest has something terrestrial jungles lack – grace. Better still, unlike the congested gloom of the jungle, this spacious grove allows you to see its beauty. It is almost as if it expects to be admired.

The forest attracts a myriad of bizarre creatures, who come to browse and hide among the fronds. Charles Darwin was mightily impressed when he first saw the forests of giant kelp from the rail of the *Beagle*:

> The number of living creatures whose existence intimately depends on the kelp is wonderful . . . I can only compare these great aquatic forests with the terrestrial ones in the inter-tropical regions. Yet if in any country a forest was destroyed; I do not believe nearly so many species of animal would perish as would here, from the destruction of the kelp.

He was right; a single square kilometre of Californian kelp forest can house over 67,000 animals. He also predicted that: 'A great volume might be written, describing the inhabitants of one of these beds of seaweed.' Eventually, that story was indeed told, here in California, by two of America's most famous divers, Conrad Limbaugh and Wheeler North.

Limbaugh, Connie to his friends, was supple and stocky and balding from youth. You might have mistaken him for a beach bum as he invariably went barefoot or, on more formal occasions, wore flip-flops, continually stepping on the bottoms of his baggy trousers. He was described a being a hippie before hippies were invented

Limbaugh first went diving in 1937 when he was only twelve, and for his original compressed air dives used one of the first two aqualungs brought to the United States by the US Navy. By that time he was a research student at the University of California's Scripps Institution of Oceanography at La Jolla, one of the world's great marine research laboratories. In 1950, he became the institution's chief diver and its first safety officer. When a student at another campus was drowned, Limbaugh devised a research diving course at Scripps, the first for civilians in the United States.

By 1955, 25,000 aqualungs had been sold worldwide and over 80 per cent of these had been purchased in California. Scripps now had twenty-five regular diving researchers, more than any other marine lab in the world.

Connie was tireless and nothing was too much trouble. A curator from a local public aquarium made an annual trip to the institution to replenish stocks. He was packing live specimens on to a truck when Limbaugh trudged by, fresh from the sea, his lips blue and skin as 'shrivelled as a Fresno prune'. He asked politely how the aquarist's collecting had gone. 'I got everything except for a couple of horn sharks' was the reply. 'Really,' said Limbaugh. 'Just hang on a minute.'

He dived straight back into the ocean and half an hour later emerged from the surf carrying a horn shark in each hand.

Many rare species and even a few previously unknown to science were also being brought out of the sea, but they were not what interested him. He devoted himself to the meticulous long-term observation of the habits of the common and ecologically important fish. It took twelve years to complete his studies of one species, the black croaker, and the results of hundreds of dives were summarised in an article only eleven pages long.

Limbaugh's powers of observation and recall were legendary. On a dive he invariably noticed everything that his buddy overlooked. On surfacing he would decamp directly to a typewriter and hammer out fifteen or twenty pages of detailed descriptions of everything interesting he had seen. After one diving trip to the offshore Channel Islands he and his buddies landed and came across an old Indian village abandoned perhaps a century earlier. They mooched around searching for artefacts. His eight companions discovered only half a broken arrowhead between them, whereas Connie found five complete arrowheads, one in the footprint of the man immediately ahead of him.

There were occasional underwater emergencies, as when an elephant seal pup gnawed his bald head, but Connie was an excellent buddy to have with you in a tight situation. Once, when burdened with gear too heavy to haul to the surface, he immediately tied it to collecting bags (intended for specimens), filled them with air and the gear rose unaided. Another time he surfaced to find that the recovery boat was far beyond hailing distance, so he got the party to ditch their weight belts and he inflated their rubber suits with air from his tanks so that they couldn't sink. It took them two hours to make it back to the beach.

Only 350 metres from the end of the institution's pier is the shallow (ninety metres) end of a narrow underwater defile called

Scripps Canyon, which he made into a submerged laboratory. He monitored the conditions and marked the lower limits of the major zone-forming species to see if they changed seasonally or from year to year. It was here that he and a companion were the first to see and film the mass mating of squid. Millions of them arrived from the ocean at the same time each year. They hovered and writhed together in pairs, the males' tentacles turning scarlet with passion, while clutching the wan females. Afterwards, they all died of exhaustion and in came the predatory sharks and bat rays to tidy up. The floor of the gully as far as Connie could swim was covered almost a metre deep with what could be mistaken for condoms. But each of these capsules enclosed eggs – hundreds of tiny squid-to-be.

Limbaugh now began a five-year study of the fish of the giant kelp forest. Never has fish-watching been done so well. It resulted in one of the most comprehensive accounts ever published of direct observation of an underwater community. Perhaps the habits of the garibaldi (*Hypsypops rubicundus* – ruddy high-eyes) illustrate why anyone would study the habits of fish.

There is more than a hint of Hollywood in the garibaldi. It flits around wearing a dazzling orange outfit with golden lips and fins, and silver rims around its eyes. Garibaldis were too gaudy for their own good and were avidly collected until designated as the State Fish of California and protected.

If a female garibaldi swims by, the male repeatedly loops the loop chuckling like a hen until she wanders closer. Pugnacious and fiercely territorial, he guards the nest of brilliant-yellow eggs against all comers and will even butt a diver who gets too close. Yet when not attacking rivals, he is a gentle gardener assiduously weeding the nest and removing everything except his favourite silky red alga.

During his surveys, Limbaugh also became intrigued by the sight of fish and shrimps picking parasites from larger fish; he even observed a seagull cleaning a sunfish lolling at the surface. The underwater explorer William Beebe had been the first to observe this sort of behaviour in 1927. He saw a parrotfish upend itself with its mouth open and then 'a school of little wrasse darted out and thoroughly cleaned its cheeks, lips, teeth and scales . . . the parrot fish remaining quite motionless all the while. It was an aquatic parallel of crocodile and plover, cattle and egret, rhino and tick bird.'

Limbaugh launched into the systematic observation of 'cleaner' organisms in different parts of the world. He discovered that many cleaners occupy specific cleaning stations and advertise their services. Like hookers, they solicited for custom by parading at the station: cleaner fish often do a little zigzag dance and cleaner shrimps adopt a distinctive 'I am ready to inspect you' position. Pederson's shrimp perches on a sea anemone and, from this conspicuous kiosk, whips its overlong antennae and sways its body as if practising the rumba. They sometimes have more striking coloration or patterning than their non-cleaning relatives, but this may be to warn their customers that they are unpalatable (even if it's a lie).

To ensure there is no misunderstanding the fishy customers also adopt an unnatural pose with the head up or down and fins outstretched. Some even change colour; black surgeonfish turn bright blue and white and yellow dappled goatfish blush red. Then they seem to fall into a trance and submit to the cleaners taking liberties in their mouth or inside their gills, even opening the gill covers to allow access, until with a twitch they announce that the appointment is at an end. Other patrons wait impatiently to be attended to and get tetchy as the queue lengthens, even though some cleaner fish treat fifty patients an hour. The crowd is sometimes so numerous as to obscure the cleaner, or may keep him prisoner until everyone has been serviced. Although opportunistic cleaners are at risk, the species of specialised cleaners seem to have immunity from being eaten, at least while at the grooming station. Fish disturbed by a diver often spit out the cleaners before fleeing.

The cleaner begins its attentions by caressing the client who may not attend just to be relieved of irritating parasites, but because it craves the stroking and tickling that the cleaner provides while going about its business. Some fish have been seen to adopt the 'grooming' pose when accidentally brushed by a sea fan.

The cleaners are not just parasite pickers, they also clean out necrotic tissue from wounds, just like medicinal maggots. Wounded fish spend almost half their time at cleaning stations and even species that rarely attend for routine grooming will turn up if wounded. Some cleaners cheat by feeding on energy-rich mucus that is easier to find than parasites. But the clients soon twig what's going on and shun a station where they are getting poor service.

Plants have similar problems and seek similar solutions. Seaweeds become infested with smaller algae that cut out the light and pinch the nutrients before they reach the host plant beneath. They rely on snails or small crustaceans to browse off the offending pests. Some seaweeds form low crusts that spread across the rock. By replacing the rock surface with their own tissue they are almost inviting settling algae to foul them. So they secrete chemicals that attract the larvae of herbivorous limpets such as abalone, which then, like beneficent barbers, keep them stubble-free. The seaweed often fortifies its crust with calcium so the barbers don't damage them too.

With Wheeler North and a couple of other friends Connie set up a part-time environmental consultancy, which made so much profit that together they opened one of the very first dive shops.

By now, Connie's diving log totalled not hours spent underwater but months, and he had written the first scientific diving manual. He was the most famous diver in America.

In 1960, he was the US representative at a meeting of the International Underwater Confederation in Barcelona. After the conference he went diving on the French Riviera. As his trip was coming to an end, he returned all the diving gear he had borrowed from local divers, but the very next day he was persuaded to have one last dive. So it proved.

It was to be a brief excursion into the caves that riddled the limestone cliffs at Port Miou near Cassis. His guide was a local, familiar with the caves. It was a short swim into the first chamber and back, a sortie sometimes used as a novel training dive for escorted amateurs. The caves had been formed by the action of a freshwater stream that still flowed out along the roof of the entrance tunnel, on top of the sea water sliding in below. In the salt water the beams of their torches penetrated perhaps four metres, but in the turbid fresh water visibility was reduced to a few centimetres. The first chamber had a distinctive six-metre-tall cairn of stones and above it was a twenty-metre chimney, like a well opening on the top of a cliff.

They surfaced at the bottom of the chimney and Connie tried to take movies looking up towards the sky, but he was over-weighted and couldn't hold himself out of the water. His buddy offered to help, but first had to disencumber himself. He dived down and placed his large torch on the top of the cairn and returned. With two hands free he was able to hold Limbaugh up to take the shot. He then hand-signalled Limbaugh to stay put while he retrieved his torch. Connie signed 'OK'. But when the buddy returned he was gone.

He must have misunderstood, followed him down and they had missed each other in the murky water. A search of the chamber below proved fruitless. Perhaps Limbaugh had made his own way out of the cave? It was only a five-minute swim to the open sea.

The exit wasn't difficult to identify as that was where the stream exited and, unlike all the other tunnels, its walls were coated with marine growths. But the problem was not recognising the exit, but finding it. The first chamber was a circle of confusion in which it was easy to become disorientated, and in all directions lay a maze of interconnecting tunnels and dead ends. The locals called it a submerged Swiss cheese.

He searched until his air was exhausted and if his was gone, so was Connie's. There was just a faint hope that he might have found an air pocket trapped in the roof of a tunnel.

Cousteau's dive team rushed to the rescue and four days later Wheeler North, Connie's friend and diving companion, arrived from the US. He was taken into the caves and recalls how the dark walls of the tunnels closed in about him. Spikes seem to lurch out from the ceiling and floor as if he were in a 'gigantic iron maiden'. Every time he shone his torch on the ceiling there were small bubbles of air trapped in the concavities. They stared down at him like luminous eyes. This was no place for human beings and no place for his friend to be lost in the darkness.

A week after he disappeared, almost to the minute, the torch beam of the search divers caught a glint of yellow. They were air tanks. Limbaugh lay slightly hunched on his side with his mouth-piece still between his teeth. He was deep inside the cave in the opposite direction from the exit. Connie was only thirty-five years old.

The post-mortem concluded he had asphyxiated after becoming lost in the labyrinth. To his friends it seemed incomprehensible that such an experienced and cool-headed diver, who had written the first manual on diving safety, should have wandered away on his own when all he had to do was to return to the base of the chimney, breathe the air and wait to be reunited with his buddy. But he didn't.

Two months later Cousteau went to the United States on business and made a detour to take Wheeler North and Connie's widow, Nan, to dinner. His sympathy was overwhelming and it was clear that even he felt that diving had lost one of its giants.

Freewheeling

Southern California

The first research diver in California was not even an American. Cheng Kui Tseng was a young Chinese botanist who, after getting his PhD, won a fellowship to study at the Scripps Institution where Limbaugh would later work. On the adjacent coast he found aquatic prairies of potentially useful seaweeds and was distressed that they were allowed just to grow and then rot away unused. He concluded that Americans were 'too rich to bother with these weeds of the sea'. But it was all to change.

The Japanese attack on Pearl Harbor not only delayed his return to China it also elevated seaweeds to a priority resource. All supplies of seaweed extracts from Japan ceased. The authorities panicked, for seaweed gels were needed for culturing antibiotics and curbing bleeding, not to mention other American priorities such as stiffening the head on beer and preventing ice crystals forming in ice cream. It was a national emergency and Tseng was the expert on hand.

By the time he returned to China in 1947, he was an authority on the exploitation of algae and as a result of his efforts China now farms everything from oysters to kelp in the sea, and does so on a vast scale. Only 26 scallops survived out of a shipment imported into China in 1982. Now every year it produces a million tonnes from their descendants.

Some scientists believe that there is 'real' science, knowledge for its own sake, and an inferior product called applied research. I see nothing disreputable in research that attempts to meet the needs of society; indeed, there is a view that most research is either applied or yet to be applied.

Several distinguished biologists have studied the processes that underpin the exploitation of fish. Without seafood the world would starve, for the majority of people get their protein from the sea. We expect to haul fish and shellfish from the ocean and do it in a big way; every day a factory ship is at sea it freezes or salts down over 300 tonnes of fish, and then manufactures 150 tonnes of fish meal, five tonnes of fish oil, and twenty tonnes of ice. On whatever coast you tread there are sea farms for salmon or milkfish or shrimps, but the scale on which seaweeds are harvested is perhaps even more surprising. A small red seaweed called nori is much esteemed in the East as human food and now we too get to taste it as the wrapping for sushi. I have seen thousands of nori-covered floating nets, each bigger than a tennis court, laid out on the sea like the street plan of a great city. The annual crop of nori cultivated in Japan alone has a value of $1 billion.

Even kelp is grown on a large scale for food, but demand in the West grew considerably when commercial companies began to exploit algin, the jelly that fills the interior of the kelp and gives the stems their flexibility. It has an astonishing ability to absorb water without getting wet and has the power to organise water molecules, which makes it invaluable if anything needs to be suspended, gelled, stabilised or thickened. Every day you eat something containing seaweed extracts. Algin is what keeps the world's ketchup flowing from the bottle and toothpaste from the tube. Seaweeds became big business and as giant kelp was so easy to gather from the surface, it became the main source of algin. Soon it was being collected by huge floating lawnmowers that cut great swathes through the forest canopy.

By the 1950s, it was apparent that there had been a substantial decline in all the kelp beds. Some that once yielded 75,000 tonnes a year had all but vanished. Was it the result of over-harvesting or pollution? In 1956, a kelp-gathering company thought an ecological study would supply the answer. They dangled a large research grant in front of the Scripps Institution, whose director desperately searched around for a biologist to lead the investigation. His eye alighted on Wheeler North who was summoned to the director's office where he protested his unsuitability: 'But I'm a biochemist not an ecologist.' The boss would have none of it: 'Wheeler, you're the ideal man for the job.'

Wheeler had taken Connie's diver-training course, but didn't look athletic; he stooped and walked with a limp as a result of a bizarre accident. His obsessive accumulation of redundant gear out-grew his storage space so he tunnelled into the sandstone cliff behind his house to make more room. The cliff face collapsed flinging him to the beach ten metres below. He couldn't move and the tide was coming in. Wheeler tucked a note into the collar of his pet dog and the resourceful Benny returned with helpers. It was a scene from *Lassie's Rescue Rangers*, but Wheeler's back was permanently damaged.

Just as penguins are awkward on land yet agile beneath the waves, it was underwater that Wheeler found fluency of movement. A collaborator recalls diving with him when a heavy swell was running. While he was being buffeted about, Wheeler 'gracefully swooped by, writing busily on his slate'.

The image of aquaman was marred by the cascade of loose straps and gauges flapping around him. He rarely bothered to buckle up the straps for his aqualung harness and all his gear threatened to be deciduous, but he was too engrossed to notice. Then there was the time his antique rubber suit split to reveal he had forgotten to wear bathing trunks.

New dive buddies feared the worst and were astonished that he was competent underwater. They were even more surprised if weeks later they asked him about a particular dive, for he could immediately recall the depth, visibility, weather, bottom temperature and everything he saw.

They had to accept that Wheeler always dived alone, no matter how many buddies accompanied him. Underwater he would rush off to get the job done and you were supposed to do the same and meet up again sometime after the dive was over. When he dived solo the task of the attendant in the boat above was impossible. Wheeler would give only the most vague outline of his proposed route and then immediately depart from it. He christened his boat the *Great Green Urinal* as it incorporated 'every mistake in boat building over the last two hundred years' and set off to assess the state of the beds in southern California.

In Wheeler's words, 'We ran our tiny skiff into the teeth of gales to measure the force exerted on kelp holdfasts by passing swells, but we could not determine what was causing the kelp's decline.'

The relentless human colonisation of California had caused small coastal towns to merge into conurbations. Much of the sewage was voided directly into the sea and there was very little kelp in the vicinity of outfalls, but in the laboratory the nutrients in the effluent were found to stimulate growth not retard it.

In 1957, when the tanker *Tampico Maru* was disembowelled by a reef south of Ensenada in Mexico and its cargo of diesel oil bled away, they rushed south to assess the damage to the local wildlife. It was severe; even some of the more hardy animals such as the urchins had been wiped out. But the juvenile kelps were unaffected and rapidly grew into a new forest. They were thriving because of the absence of urchins.

Sea urchins were ubiquitous on the California coast with up to eighty individuals per square metre in places. And they didn't have to consume a whole plant of giant kelp, let alone an entire forest, for merely nibbling the base of the stems enabled the next storm to snap them off and carry them away.

Wheeler's team observed advancing fronts of urchin armies demolishing the kelp and leaving behind what is now called an urchin barren. Urchins were the locusts of the forest. Normally they were lazy locusts and, if enough detached blades drifted down from the canopy, they would stay at home to dine on them. But sewage effluent had encouraged the urchins to burgeon and kelp harvesting meant there was no longer sufficient seaweed brought to their door, so they went on the rampage.

In the old days there had been fewer urchins around because they had a voracious enemy. The cuteness of the frisky, whiskery sea otters is lost on the clams and sea urchins they devour. To make ends meet, an otter must consume 25 per cent of its body weight every day. There had once been tens of thousands living along the California coast, but at the beginning of the twentieth century they were hunted almost to extinction for their pelts. In their absence the urchins multiplied. When otters were reintroduced to Monterey Bay, the urchins suffered and the kelp blossomed.

The team also tried to determine whether harvesting reduced the diversity of the forest-dwelling animals. Over fifty types of fish lived among the kelps, but Wheeler decided that the noisy bubbles escaping from his regulator were scaring some fish away, so he raided his store for a wartime oxygen re-breather that didn't liberate bubbles. The canister was rusty and the bladders half perished, but no matter. His minder on the surface peered anxiously into the water, expecting at any moment to see a limp body or a great belch of gas as the canister exploded. Now that *would* have scared the fish.

The group was asked to restore the kelp beds at Palos Verdes off

San Diego where the former forest had dwindled to just two plants. To make aerial surveys of the beds Wheeler bought a Piper Cherokee. He flew alone, for no one who had been a passenger in his car would accompany him into the air. To take vertical shots a hole was cut in the floor to push the camera lens through. The procedure was simplicity itself; he put the plane into a slow bank, then left the cockpit to take a few snaps before resuming control just in the nick of time.

Wheeler enlisted 250 amateur divers to remove urchins. Meanwhile, his team towed in 2,000 healthy plants from seventy kilometres away and bred millions of tiny plants to swamp the area. Soon the Palos Verdes forest was restored to its former size and was discarding sufficient old blades to keep the remaining urchins replete and at bay. The kelp forests of California were on the mend.

There was an oil crisis. Motorists were getting restless in the long queues at the gas station and started shooting each other. I knew someone who had an additional petrol tank fitted to his car to double its range.

The search was on for renewable sources of energy. One idea was to produce methane gas from fast-growing plants. The Gas Research Institute believed giant kelp was the answer and enlisted a sceptical Wheeler to be farmer-in-chief. He had already decided that the best way to grow kelp was on large cat's cradle of rope, but the funding was for a much more imposing structure: the Offshore Test Platform.

Our boat headed straight out to sea from the coast and took over an hour to reach the platform. In the unending expanse of ocean even the platform's huge buoy looked lost. It was as big as a farmer's silo above the surface and went down a further twenty metres underwater. From its base radiated eight thirty-metre-long horizontal arms like the spokes of an umbrella. Cables strung between the arms were festooned with young kelp plants.

The principle was simple. Kelps grow rapidly if given plenty of nutrients. The surface waters are often low in nutrients, but the sea floor, where every dead organism sinks to decompose, is an inexhaustible warehouse of unused fertiliser. All they had to do was to retrieve it.

Twisting around the buoy like giant anacondas were three hoses, each over sixty centimetres in diameter, snaking down until they were lost in the haze. Three immense diesel pumps sucked up nutrient-rich water from 450 metres below and sprayed 630 litres per second over the plants. And that is *exactly* what they did – the water went *over* the plants and the currents swept the nutrients away.

I dived down below the platform into the heart of the ocean so that I could turn and see the structure silhouetted against the sky. The sea floor was still six hundred metres below although the surface now seemed far, far away. The huge serpentine hoses reaching from the tiny disc above and vanishing into the nothingness below only enhanced the eerie unreality of being suspended nowhere. Not far below here the diver may enter a strange place where the surface can no longer be seen and the bottom is not yet visible. It is a blue void, featureless, without dimensions. Only the ascending quicksilver bubbles you exhale seem to know where they are going. It is a place where the mind gets lost.

I was relieved to return to the platform. The whole structure was in motion and after the deathly loneliness below, it seemed to be alive, with every sinew tensing and relaxing in the swell. It creaked from every joint and from somewhere deep inside it moaned like a dinosaur that knew it was doomed.

The delicate kelps were also doomed, for as the platform rose they were still descending and they abraded against the cables. Within a couple of months they would all be destroyed. One night the huge test platform vanished and was never seen again. It had cost $9 million. Some say it died of embarrassment.

On the way home we called in to Wheeler's lab at Corona del Mar. A kindly and modest man, he referred to himself as a scuba-forester, a profession that failed to endear him to locals who considered kelp to be God's contrivance for fouling propellers or infusing the beach with the sweet smell of decay.

I fell into an easy camaraderie with Wheeler as I had with Mike Neushul. Marine biologists are a relatively small community. Even as a student I knew by name many of those in my chosen field. I had read their articles as one would the latest short story by a favourite author. When we met for the first time, we had an immediate rapport because we shared obsessions and a mutual agreement on how problems could be solved.

In 1982 the Falkland Islands were invaded by Argentina and British troops were dispatched. While they were en route, I received a telephone call from the Ministry of Defence.

'Are you the seaweed chappie?' said a man with a pound of plums in his mouth. 'Just a wee enquiry. I've been led to believe there are exceptionally large seaweeds off the coast of the erm . . . Falkland Islands.'

'That's right,' I replied. 'Giant kelp is very abundant there.'

'But only in deep water, I presume.'

'Well, yes, but it forms a dense canopy on the surface.'

There was a pause. 'The leaves and things, don't suppose they could impede the progress of small craft coming ashore?'

'The stems might tangle in the propellers.'

'Oh,' he said and then there was such a long silence I felt obliged to offer reassurance. 'It's mostly an open-coast plant. There would be much less of it in sheltered coves and inlets.'

'Really, by Jove. Is that so . . .?'

I don't remember the conversation being much longer than that, but I have occasionally wondered whether it was seaweed, or the lack of it, that determined the landing place of the British troops.

Moonstruck

Santa Barbara, California

Seaweeds are oversexed. *Sargassum muticum* devotes half its body weight to its reproductive organs. In season it produces 35,000 eggs for every gram of its tissue. It doesn't release the eggs as soon as they are ripe. Instead, it holds them back and then extrudes them in pulses every fourteen days or so. Masses of them perch on the outside of the reproductive organs waiting for something to happen. That something is fertilisation.

Millions of sperm are liberated on the same day and set off in search of the eggs. In the chaos of the sea it is surprising that any of them make it. Some seaweeds only release on calm days when the sperm are less likely to be washed away, although how they perceive whether it's choppy or not is a mystery. *Sargassum* can't wait. Its sex cells pop out every fortnight whether it is stormy or still. Fortunately, the sperms are not navigating blind, for the eggs give off an irresistible odour from chemicals called pheromones. The sperm cells are attracted by light so they probably swim upwards and water movement swills them sideways towards adjacent plants. When they are close enough to get a whiff of the egg, they home in. Under the microscope you can see hundreds, even thousands of sperms frantically swarming around a single egg. The first one to break in stimulates the egg to immediately erect a chemical barrier to the others, for multiple fertilisation is lethal.

Pheromones that attract a mate are also found in animals. Indeed, we sometimes use them in perfumes. And we humans have our own natural supply. The pioneer sexologist, Havelock Ellis, told a story of detecting a slight odour from the man sitting next to him. He found it unpleasant and unsettling. When the man left, Ellis realised that the scent was actually emanating from a young woman who had been sitting just beyond the man – and it was really quite pleasant. These chemical signals work their magic without us consciously perceiving them and may have dramatic consequences. Women living together in a nurses' residence were found to have synchronised their menstrual cycles, presumably having been orchestrated by whoever was emitting the strongest pheromones.

Producing sperm and eggs simultaneously in large batches as *Sargassum* does increases the likelihood that they will meet. But what stimulates it to do so every fourteen days?

There is a tiny tide within us all. Hormones ebb and flow in our bodies to a daily rhythm. A man's libido peaks around three in the morning, which is also when he is most prone to a heart attack. Our temperature, pulse and blood pressure rise and fall in time with the spin of the Earth.

Our natural rhythm is not that of the twenty-four-hour day, but almost fifty minutes longer, the time it takes the moon to make a single orbit around the Earth. People kept in a windowless room with no clues as to the time ease naturally into the longer cycle. We are all lunatics, slaves to the moon. Some of our hairier friends may become werewolves when the moon is full, others will merely get arrested for disorderly behaviour, for there appears to be a peak of binge drinking every twenty-eight days – a lunar month. Every full moon cops, firemen and midwives in New York brace themselves for a busy night.

It is not surprising that the behaviour of animals living on the shore is often governed by the tides, whose rise and fall are determined largely by the position of the moon. Young crabs that have to moult their rigid shells in order to grow, do so most commonly when the highest tides occur, every fourteen days.

Many internal biological rhythms are conditioned to coincide with the tides. They may persist for a time in the laboratory without benefit of the ocean. But marine creatures have felt the tides for millions of years and many of them respond directly to them. It is obvious even to a limpet whether it is in air or water and, from the direction of the flow, whether the tide is arriving or departing. It may also be possible to gauge by the weight of water above whether it is a really high tide or a lesser one. Crabs collected from a non-tidal harbour exhibited the same rhythm of activity as those living on the adjacent tidal shore. They were following the ebb and flow of the moon itself rather than its slave, the tide.

When Spanish explorers first came to California the local Indians told them that fish came up and danced on the beach – and they were right. Three nights after the moon is full or new, Santa Barbarians take a nocturnal barbecue on the beach. They don't bother to go fishing; the fish come to them.

I came to see for myself. The tide came to flood, lingered a little and then, as it began to recede, little fish called grunions swarmed ashore. Within minutes the sand was lost beneath a slither of glittering silversides. Sometimes there are dozens of fires on the beach for a bumper grunion grill, but that night everyone seemed content just to watch the show. And what a show it is.

There were thousands of sparkling writhing bodies. Their glum fishy look didn't fool me; they were as excited as fish can get. The female wriggled tail first into the sand until it was almost buried and laid 2,000 eggs below the surface. Meanwhile, the male coiled around her and did what males do. There were as many as eight males to each female. For some species such as sea horses and pipefish, courtship is a stately dance, a marine minuet. But for grunions it's an orgy.

Within forty minutes they had all returned to the sea and the sand was deserted. The water had receded and the tides would not reach this high again until the spring tides recurred in a fortnight's time at the next new or full moon. By that time the eggs would have incubated in the warm damp sand and as soon as they were stirred by the waves they'd hatch. Tiny fishlets would pop up in their thousands and flitter away into the surf. They would not return for a year, when the moon calls again, and then they would lay their eggs for the first time.

Every fourteen days when the moon is right the mating ritual is repeated by grunions oblivious of Shakespeare's warning:

> Oh! swear not by the moon, the inconstant moon,
> That monthly changes in her circled orbit,
> Lest that thy love prove likewise variable.

But the moon is not inconstant, it is as predictable as a parson's sermon. I have on my computer the timing and height of all the high and low tides every day until the year 2050, based on the positions that the moon and sun will occupy in the heavens on those days. Now that is predictability.

The rhythm of the moon orchestrates many ripenings and releases of spores and eggs in readiness for brief encounters. Oysters liberate their larvae eight to ten days after the full and new moon, but some of the most spectacular lunar rhythms are shown by worms. The palolo worm in Samoa swarms seasonally at the surface, but always at the same phase of the moon. The date of the swarming is so predictable that it is marked on Samoan calendars as 'Worm Day'. Worms are gathered by the locals for gourmet eating. Some impetuous epicures swallow them alive like strands of animated spaghetti.

You might think that for those species living on the dim sea floor the tides are less important, but still they respond to the lunar pulse. Some species of underwater worm spend thirty days packing a special new back end with eggs and sperm so that they are ready for release just before the full moon. When ripe, the rear end detaches like a space module and swims to the surface leaving the real worm below on the bottom. On a single orgiastic evening they gather in their thousands churning the surface waters into what has been called a 'frothy brothel'. When the eggs are fertilised the modules are finished, but the original worms deep below are already at work packing a new module for the next full moon.

They are responding to the duration of the night. In the laboratory two to four nights of simulated 'moonlight' regiments their reproduction to a monthly cycle so that the sex cells are well met by moonlight.

Sardines and Scarlatti

Monterey, California

Philip Henry Gosse, who boasted he had never read Shakespeare and refused to tolerate novels because they gave 'false and disturbing pictures of life', would have been dismayed to learn that he is best remembered as the harsh father in Edmund Gosse's *Father and Son*, that he inspired Charles Kingsley to write *The Water Babies*, and that his life contributed to Ann Lingard's *Seaside Pleasures*, and Peter Carey's Booker prize-winning *Oscar and Lucinda*. I can think of only one marine biologist who has inspired better literature. His name is Ed Ricketts and he has been a hero of mine since my teens.

Ricketts took courses part-time at the University of Chicago. He stoked boilers first thing in the morning, attended classes during the day, then went back to the furnaces for some more stoking before serving in a store during the evening. Ed also fell for a woman whose husband worked nights, so he didn't recall getting any sleep at all. He went all the way with the woman, but failed to complete the university course and left after a year. After drifting between jobs he went to work for a firm that supplied biological specimens to schools and colleges. In 1923, he took his expertise and the company's mailing list to Monterey in California and set up the Pacific Biological Laboratory.

Ed was the least businesslike trader in the world. His specimen supply trading was a shambles. Orders poured in, piled up on his desk until they avalanched on to the floor and were eventually thrown out. When his nose told him that a parcel had begun to decompose, he opened it to find an unidentifiable suppurating mess. Only a note inside revealed that long ago it had been a cheesecake. While ignoring customers for much of the time, he nonetheless sent out catalogues periodically to solicit further orders.

One catalogue advertised 'quantities of delightful and beautiful hagfish'. To others it might win the vote for the most repellent creature alive, whose most endearing accomplishment is to instantly turn a bucket of water into solid slime, but Ed probably did find that delightful.

Mostly, he supplied creatures fixed in formalin and must have embalmed a heap of cats, dogs and dogfish. He had no qualms about killing animals, but despised cruelty. He would tip his hat to dogs as he passed and once, on seeing a stranger whacking a hound with a broom handle, he leapt from his car and gave the man a good hiding.

He was small and slight, but amazingly strong and blessed with everlasting stamina. His clothes were held together with safety pins and open to the air at knees and elbows. Few of the garments appeared to be in his size.

Ed loved and, if possible, made love to every woman he met, but he drew the line at those with thin lips, unless they drew their lipstick line boldly to give a facsimile of fullness. He served as counsellor and agony aunt for the girls at the bordello across the street. After the usual Saturday-night fracas he also practised as a discreet surgeon.

The only time he was punctual on purpose was when he had to catch a tide. No exploration was too difficult. On one occasion he crawled right inside the fetid carcass of a dead basking shark to retrieve its liver. He was most at home on the shore exploring the gullies and pools, with a heron watching him at work and waiting its turn to fish. He was captivated by the shelled and slithery pool dwellers. There were no trivial motives here, every creature was in earnest; they either ate or were eaten, they must fight or flee. Ricketts believed that human behaviour could be illuminated by close inspection of the communities of a pool.

If he didn't know the name of a creature then it was a good bet that nobody else did either. Ed never failed to sniff every animal and plant he collected and could probably have identified them in the dark. He also tasted quite a few — and not just the edible ones. Wondering why a sea slug seemed to have no predators, he put it in his mouth and found out — within seconds he was retching. The slug was bright orange to advertise its acrid taste to would-be consumers, but Ed had failed to read the signs. After tasting a sea anemone to test how fierce its sting was, he couldn't close his lips for days.

While housewives bottled fruit, Ricketts placed animals in bottles fragrant with formalin where they would slowly stiffen and bleach. One of his preserved specimens became well known locally and attracted the curious. It was a well-developed foetus with an oriental look. It sat cross-legged in its jar like a small ivory Buddha. A woman came in to see it and then in exchange undressed to show him her Caesarean scar.

Ed wasn't content just to pickle and parcel marine creatures, he was set on writing a book on the distribution of the marine fauna of the Pacific coast and enlisted Jack Calvin, a scribbler of Sunday school stories, to help smooth his writing. Although he knew nothing about biology, Calvin would be co-author.

It was not to be the usual annotated list of species in taxonomic order. The book would be arranged according to habitat, with sections on the open coast, bays and estuaries, wharf pilings and so on, and within each section there were to be chapters on the inhabitants of different zones, and those on rock as compared to sand. And in the telling it grew from a beginner's little pocket guide to a substantial tome.

He sent the manuscript to a professional zoologist who didn't like the arrangement and positively hated the idea of a book that purported to be useful to the professional marine biologist yet intelligible to the general reader.

In those days there seemed to be a widespread belief that for a fact to be valuable, it must also be dull, and if it were irresistibly interesting, the author must strive his utmost to render it tedious, thereby enhancing its air of authenticity. Ricketts and Calvin's book had a charming informal tone: 'The spiny lobster which often meets the fate of the innocent oysters that accompanied the Walrus and the Carpenter . . . unless captured with a spear or trap, has a good chance to evade the path that leads to mayonnaise and seasoning.' Anthropomorphic asides about hermit crabs that were busy lovemaking or fighting with enthusiasm were red rags to the bullish professor, whom Ricketts christened 'Old Jingle Bollicks'.

In spite of him, *Between Pacific Tides* was published and deservedly became the bible for anyone studying shore biology on the Pacific coast. Never has a textbook been so readable. It overflows with Ricketts's joy of discovery and his celebration of the variety and abundance of marine life. Although it has been academicised a little over the years by editors who were professional marine biologists, nothing could extinguish its original sparkle. Ricketts found colour and life everywhere on the shore, and this is reflected in his book. John Steinbeck caught its value in his foreword. It was not merely intended to provide information, but to stimulate curiosity. He also held out an irresistible invitation: if the readers kept their eyes open they might discover something that the authors had overlooked. After all, all knowledge is incomplete.

In 1930, Steinbeck had been sitting in a dentist's waiting room when Ricketts emerged from the surgery, bloodied and clutching a large molar complete with a piece of his jawbone. This seemed worthy of comment and the two struck up a conversation and an almost instant friendship. The twenty-eight-year-old Steinbeck had just come to live locally. His first novel had been published, but *Cup of Gold* was, by his own admission, an amateur effort that hadn't repaid his meagre advance of $250. The two men had much in common; they were close in age (Ricketts was older by five years), both had dropped out of university, worked as surveyors' assistants and were seeking a philosophy of life. More surprisingly, they were both attempting to forge it from the unifying principles of biology, for Steinbeck had been deeply influenced by a course in marine biology he had taken at Stanford University.

This was to be the most formative time of Steinbeck's life and within nine years he had published *Tortilla Flat* and *The Grapes of Wrath*, which came out in 1939 in the same month as *Between Pacific Tides*. His and Ricketts' reputations were established simultaneously.

Steinbeck joined the group of freethinking, hard-drinking writers, amateur philosophers and professional layabouts that congregated at the laboratory. Ed always had an incurable acceptance of mystics, if they could hold their liquor. Wine could be had for only thirty-nine cents a gallon, beer milkshakes were as good as food, and the hooch affectionately known as 'old tennis shoes' was cheap and effective, so long as you ignored the taste and didn't mind the temporary numbness of the lips.

The parties sometimes lasted for days, and were joyous for the opening forty-eight hours or so. If by the second night the boys were getting maudlin, Ed would lift their souls by reading from Goethe's *Faust* and playing some Scarlatti or Palestrina or, best of all, Bach.

The lab seemed an unpromising setting for either debauchery or Bach. Its shelves were lined with aquarium tanks with air bubbles chuckling through them to keep the crabs and anemones alive. There were also dozens of cages full of scuttling rats, lazy lizards and rattlesnakes. One day a woman came in and put money down to pay for a rattlesnake, insisting that it must be male. She would leave it at the lab and her only requirement was that she should be present at feeding time. She asked for it to be fed immediately. As the snake slowly engulfed a rat, she watched intently and her mouth opened wide as if she too were at the dining table. As the rat inched slowly into the snake, her jaws moved in synchrony with those of the reptile. Although she paid for a year's supply of rats, she never returned to watch them disappear – much to Ed's relief. Steinbeck turned the incident into a short story, which was considered too grotesque and far-fetched by the critics. How little they knew of humankind.

Ricketts' Pacific Biological Laboratory was situated on Ocean View Avenue in Monterey, known to everyone as Cannery Row. The small wooden lab huddled between the Monterey Canning Company and the San Carlos Canning Co.; next door to that was Portola Brand Pilchards. The first fish canning factory had been built on Ocean View Avenue in 1902 and within a couple of years, Knute Hovden, a young Norwegian with a degree in fisheries engineering, mechanised almost every stage of the process from cleaning the fish to soldering the cans. During the First World War, the US military needed compact field rations and the demand for tinned sardines soared. As fast as the sardines were caught they were packed as tight as, well, sardines in a can. The more sardines the greater the profit, for the olive oil around them was more expensive than the fish. From 1915 to 1918, production went from 75,000 to 1,400,000 cases a year and the price trebled. By the 1930s, canneries lined the seaward side of

Cannery Row so that the fishing boats could moor alongside and pour their catch directly into the factories in a continuous silver cascade.

There were concerns that the rapid expansion of the fishery and the vast numbers being caught might not only strain the sardine population, but also the ecological balance of local waters. As early as 1924, a fisheries scientist warned:

> The sardine is a source of food for almost all our other great fisheries, such as the albacore, barracuda, sea bass and tuna. Tampering with its abundance may result disastrously to many interests . . . Unnecessary drain on the supply should be avoided until research has shown that it is possible to detect overfishing in time, and for that reason it is [my] belief that the use of sardines for fertilizer should be emphatically condemned.

Compressing fish offal to extract the oil then baking the residue to produce chicken feed or fertiliser had originally been an economic way of getting rid of waste, but it soon became a major product of the fishery. The industry had sufficient political clout to ignore the advice of mere biologists and no limit to the annual catch was imposed. And after all, the sardine was one of the most fecund and fast-growing fish in the sea. It would be impossible to overfish it.

In 1945 Steinbeck published *Cannery Row*, which told the story of the hobos and hookers and the lecherous but lovable marine biologist, Doc. Seven of Steinbeck's books would utilise characters from the Row, but this book and the later *Sweet Thursday* were based directly on Ricketts and his neighbours.

In those days it was bustling place with more colour than a careless painter's shirt. It was a kaleidoscope of honky-tonks and whorehouses and Wing Wong's Chinese Grocery. But most of all it was a noise and a smell. All day long the trucks rattled to and fro and the baking of fish debris produced the most nauseating stench known to man.

Steinbeck's book made Ricketts a celebrity. Students came down wanting to work with him or, better still, get invited to one of his parties. Tourists gawped through the windows and some just strode in to browse around and ask questions. Ed knew how to get rid of them. One woman was curious about a sheet of tissue-like material hanging on the wall. He put her mind at ease: 'That, madam, is the foreskin of a whale.' She turned and fled.

Steinbeck's *Cannery Row* captured the essence of the place beautifully, but no sooner was it caught than it was lost for ever.

Monterey was the world's busiest fishing port after Hull and Stavanger. The sardines were inexhaustible, until one day they were gone. Within three years of the publication of *Cannery Row*, the fishery was in terminal decline and the canneries began to close.

Sardines congregate in massive schools and are therefore easy to harvest in huge quantities. The boats ventured up to a hundred miles offshore seeking the oil slicks sardines leave on the surface, and used huge purse seine nets that could encircle an entire shoal and bag it. A single trip might catch 2,000 tons of fish per boat. A change in the ocean currents which altered the water temperature was all that was needed to put paid to a stock already stressed to the limit by over-fishing. The fish never returned and the Monterey sardine industry closed down in 1952. In the old days the clatter and stink could be experienced many kilometres downwind, but now they were only a memory.

I first visited Cannery Row twenty years later. The stink had gone but the place was redolent with the ghosts of the past. The roofs of the derelict canneries rattled in the wind. Among the debris was a bale of unused labels that never got to embrace tins of Portola Brand Pilchards — *packed from the choicest fish in Monterey Bay*.

Ricketts' old lab looked little changed and there was still some of his stuff in the basement. The place was used as a sort of club where old boys gathered to carouse as best they could, and reminisce about Ed and John and the grand old days when they didn't give a damn.

And on those evenings when the sea fog eased into town and they had drunk a little too much, perhaps they imagined Ed was back among them, telling jokes and philosophising. Or maybe making them listen to a scratchy old 78 on the phonograph. His great gift was that he taught everyone around him to look and listen.

'Now listen,' he might say. 'These guys Bach and Scarlatti came closer than anyone to passing through the back of the mirror and on to a new plane.'

They would obediently hush and for a while only Scarlatti spoke. Until Ed could hold back no longer:

'Now. It's about to happen. Do you hear it? He's breaking through. Listen . . . Listen . . .'

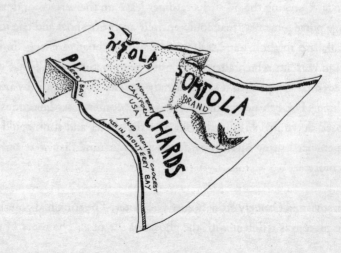

Deserts and Dreams

Sea of Cortez, Mexico

In 1979, I joined a field trip to the Gulf of California in Baja, Mexico, with John West and his graduate students from the University of California at Berkeley. John was slim with a moustache and a small goatee. He looked as if he had dashed straight from MGM where he had been auditioning for *The Three Musketeers*. Between engagements he was a botanist who studied the life histories of plants – how they got from the spore to the adult and then back again. John also taught the most popular course in the university. It was entitled 'Chocolate'.

I took with me a copy of Steinbeck and Ricketts' *Sea of Cortez*. After *Between Pacific Tides* Ricketts had planned a companion book called *The Outer Shores* that would describe the Pacific coast of Canada and Alaska, and a handbook of the marine invertebrates of San Francisco Bay. Steinbeck began to write the text from Ed's notes. Neither book came to fruition, but the same division of labour was to result in the delightful *Sea of Cortez*.

In the spring of 1940 Ricketts, Steinbeck and his wife Carol set off on a six-week cruise to the Gulf of California. Steinbeck preferred its old name, the Sea of Cortez, after the conqueror of Mexico who discovered it. In those days Europeans were intent on discovering places that had never been lost.

Steinbeck hired the *Western Flyer*, a twenty-three-metre purse seiner, before it was laid up at the end of the sardine season. Two of the crew were Italian Americans. They reflected the history of Monterey. In 1906, a Sicilian fisherman called Pietro Ferrante arrived in town. He decided the local fishery could be improved by some of the methods used back home and imported Sicilian fishermen and their *lampara* nets (from *lampo*, meaning lightning, as they were quick to cast and haul in). The nets stimulated the growth of the fishery. By 1913, each *lampara* boat captured twenty-five tons of sardines a night.

We had no nets; our expedition travelled light – apart from two or three crates of beer. Experience teaches you that any accident only destroys or renders inoperative things that are absolutely vital for the success of your mission. Trivial or useless items are immune to storms, fire and forgetfulness. Logic dictates, therefore, that the vulnerable essentials must be kept to a minimum and the bulk of the gear should be superfluous to requirements. Ricketts took masses of collecting gear, a small library of textbooks and several aquarium tanks with polarised glass put in the wrong way round so the fish could look out, but the biologist couldn't see in. To preserve their collections they took a fifteen-gallon barrel of alcohol, which they hoped the crew wouldn't discover. There was also a movie camera that was hardly ever used. So there would be no shots of the *Western Flyer* bracing against the whitecaps, or Ed and John wading in the tide pools and carrying back the dip nets with the sun setting behind them.

Ricketts and Steinbeck sailed down from Monterey. They took the slow wet way; we took the fast dry way through the desert. Ours was an epic tale of suffering under the blistering sun and our struggle to keep the beer cold. We had an electric cooler that plugged into the cigarette lighter on the dashboard. It was for keeping the live specimens chilled on the journey back, but beer bottles occupied two-thirds of its volume. Americans fear that if the ale becomes less than frigid, it would reveal its total absence of flavour.

The Mojave Desert bristles with air bases and missile ranges; it was at the China Lake Naval Weapons Center that Alex Charters did some of his ballistics work. On the China Lake site there are thousands of prehistoric petroglyphs that look as if they have been newly pecked from the rock. These protectorates are the best-guarded museums in the world. Elsewhere, outside the protection of military fences, there are huge intaglios half obliterated by vehicle tracks. Deserts bruise easily and the scars persist. Hereabouts there are the tracks of tanks from when Patton's army was on manoeuvres in 1940, at the exact time the *Western Flyer* was running down the coast to Baja. There are even the wheel marks from wagon trains that crossed these sands almost 150 years ago.

These deserts are a strange mixture of overuse and abandonment. There are many long-deserted mines with evocative names such as Chief of Sinners and Queen of Light, yet on the edge of this arid wilderness we passed Palm Springs, a community with 15,000 swimming pools and forty-two golf courses — one for every 2,000 residents. Palm Springs is an expensive way to spend water, whereas on the far side of the desert, in Lake Havasu City, London Bridge spans hardly any water at all.

In 1905, a man-made diversion of the Colorado River overflowed and flooded a depression fifty kilometres long that became known as the Salton Sea. It took $3 million to force the river back into its old course. For a while, before the lure of Las Vegas, the Salton Sea was *the* place to be seen. It has remained the favourite Californian hangout for flocks of egrets, cormorants and boobies feeding on the introduced fish.

Evaporation in the desert heat has made the lake saltier than the sea. Marine barnacles arrived on the hulls of runabouts towed by weekend sailors from the marinas of Los Angeles. They now live over 130 kilometres from the nearest tide.

The Salton Sea may become extinct should nature pull the plug and let the water run away. The San Andreas fault runs down the eastern side of the lake, the San Jacinto fault slides up the western shore. San Bernadino once boasted it was 'a city on the move'. It didn't lie. In the famous San Francisco earthquake, a 435-kilometre-long stretch of California shot seven metres northwards in seconds. That sort of shift could make a spectacular draining of the Salton Sea.

The sulphur-dusted Sonora Desert is home to all manner of barbed and leafless plants. They contain water and every animal knows it, so a thorny defence is essential. They have descriptive or mellifluous names such as 'ocotillo', 'organ pipe cactus' and 'teddy bear cholla'. Some of the big saguaro cacti are said to be up to two hundred years old, but they are so riddled with woodpecker nests that their time seems almost up.

At last the desert met the Gulf of California. Behind us there was an arid wasteland, on which the occasional lizard trod, but before us was a sparkling azure aquarium swarming with life. The gulf is a linear sea, over 1,200 kilometres from north to south, comfortably longer than Italy, but little more than 150 kilometres wide, and embraced by desert on three sides. It is a place of treacherous reefs and sudden squalls. There were few navigation lights to guide you safely into harbour, and Steinbeck tells of one that ran on the generator of the local cannery so it beamed brightly all day long, but when the cannery closed, the light went out for the night.

The shores were wonderfully rich. The tide's retreat uncovered great aggregations of animals: hundreds of thousands of black brittle-stars and green sea cucumbers and crisp brown seaweed balloons that crunched underfoot.

As I swam underwater through meadows of seagrass I wondered

why Ricketts, who was so enthralled by the inhabitants of the sea, had never gone diving. For a few dollars a handyman back in Monterey could have set a window into a large pail and opened Ed's eyes to the underwater world.

When scouring the shallows in the gulf he glimpsed the excitement that lurked below: the chattering of broken water from a dense shoal of fish just beneath the surface, or the explosion as a giant manta ray leapt into the air. But down below and hidden from sight on the sea floor emerald moray eels snarled from every cranny in the rock, hordes of hammerhead sharks gathered and belligerent bull sea lions were ever-ready to defend their harems against all comers. Every tooth was newly sharpened; every yawn a practice bite.

Was Ed nervous about diving? I don't think fear was in his vocabulary, but even the bravest men may have irrational phobias that cause them to shun confined spaces or a spider in the bath. Ed often waded chest-deep into the sea, but he never dipped his head beneath the water. He had a pathological aversion to getting his hair wet and donned a yellow sou'wester at the first sign of rain. He even wore it in the shower.

As the *Western Flyer* steamed between collecting stations, the crew jigged for squid, or trolled for skipjack and yellowfin tuna. If they spied a dense school of fish they rushed out with a net to catch them; when a manta foolishly swam too close they tried, but luckily failed, to harpoon it. What would they possible do with a million small fish or a manta with a four-metre wingspan? They had to try of course; they couldn't help themselves. They were fishermen.

At night, the *Western Flyer* anchored in bays where the quiet was broken only by the occasional lop of a fish leaping from the water. They drank and told tall tales until the deck became damp with dew.

Imperceptibly, after weeks on this shimmering sea Ed's expedition had become a dream divorced from time. Half a world away was a world in turmoil: Denmark and Norway had fallen to the Nazis, Holland and Belgium were braced for blitzkrieg and in Hitler's mind the invasion of Britain was imminent.

In 1979, we were without a radio and for a while the world continued to spin without us. During our absence all that happened was the nuclear reactor at Three Mile Island went into meltdown, rebels overthrew governments in Grenada and Nicaragua, the deposed prime minister of Pakistan was executed by the military regime, Vietnam invaded Cambodia and Soviet troops poured into Afghanistan.

Too soon our sojourn came to an end and we had to return to the real world. On the journey home across desert roads with no surface to speak of, the air conditioner in our minibus went on the blink and the beer began to warm up. In the distance was a train. It must have been at least a kilometre long, with a hundred heavy ore wagons hauled by five locomotives. Even with all this pulling power, it moved in slow motion. I was astonished at how slowly American trains travelled. The Pacific Coastal Express that went through Santa Barbara would have been no match for a fit cyclist. These mechanical slugs were the least dangerous trains in the world, I thought.

One April evening in 1948, Ed Ricketts packed up for the day and drove towards the shops to get something for dinner. He had decided he *must* get a new car. His ancient sedan had become more reluctant to start by the day and was so noisy in motion that pedestrians had to abandon conversations and cup their ears. As he shifted into second on a hill, the grating noise was so deafening he never heard the Del Monte Express as it came round the back of a warehouse towards the level crossing. The locomotive's cowcatcher crumpled his car as if it were paper and went on grinding it into the track for a hundred metres. I have seen a photograph of him lying on a stretcher beside the track with the gigantic locomotive towering over him.

Ed was in a bad state. He was rushed to hospital and they opened him up. They removed his spleen, but it was a token. Many of his bones were smashed and most of his organs punctured. When he recovered consciousness he asked how bad it was. 'Very bad,' the doctor replied.

To the doctor's astonishment he lived for three more days. He was so full of life he was a difficult man to kill. It took a train to do it.

Ghosts in
the Marine

Port Erin, Isle of Man

In 1983, after returning from one of my trips to the United States, I became Professor of Marine Biology at Liverpool University and the cocky student returned to haunt the staff at the Port Erin Marine Laboratory as their new director.

The laboratory had been involved in fisheries research for almost a hundred years. The fishing industry was in one of its recurrent crises. Once, we believed that the ocean's bounty was inexhaustible. In 1868, the Westminster Parliament felt so confident that it repealed all fisheries legislation and declared that 'unrestricted fishing be permitted thereafter'. This was at the very time that the fleets were burgeoning and steam was increasing the ability to haul ever larger nets. Not surprisingly, fish stocks plummeted. All over the coast newly formed local fisheries committees passed by-laws to halt the decline.

In those days herring, the 'silver darlings', were king. When Dutch fishermen poached 'our' herring, it led directly to the founding of the British Navy and to the war of 1652–4 that wrested dominance of the seas from Holland. The Russo-Japanese war of 1904–5 was also partly a tiff over herring stocks. For three hundred years herring were the major inshore fishery around the British Isles, and every tiny coastal village had a clutch of fishing boats. By the end of the eighteenth century the Isle of Man had over 340 herring boats, each with a crew of seven or eight, as well as fifty or so bigger fishing smacks of forty tons or more. The smacks were replacing the old square-sailed 'scoutes' whose suspect seaworthiness contributed to disaster in 1787 when a sudden gale surprised three hundred of them off Douglas. They scurried for shelter in the harbour, but the lighthouse had been demolished by a storm the previous winter. The temporary lamp attached to the ruins of the pier was struck down by the first boat to enter harbour and in the darkness fifty or sixty boats foundered and several hundred men may have drowned.

The market for Manx herrings extended from the Mediterranean to the West Indies, where it was the staple for slaves. But when they were emancipated in the 1830s their dietary desires widened and the fishery collapsed. By 1880 it had recovered and every harbour bristled with a thicket of masts. A fleet of six hundred boats operated from Manx ports, the piers groaned beneath the stacked barrels of salted fish and the air was pungent with smoke from the curing houses. The shipyards at Peel were busy building twenty-four fishing boats a year. Two-fifths of the Manx population worked in the fishing industry and their prosperity depended on it. 'No herring, no wedding' was the way the locals put it.

It was said that herring were the most abundant fish in the sea. But they were always an uncertain harvest. When there were gluts, the value fell to almost nothing. In 1793, herring were so plentiful they were caught by hand on the beach and sold for as little as four pence a hundred until no buyers could be found. Then they had to be carted off for manure. In between the boom years, there were catastrophic declines as in 1822 when fishermen and their families were destitute and near starvation. Yet the following year herring were so abundant that on a boat under way you could hear hordes of them colliding with the hull. They attracted fishermen from as far away as Cornwall. Eventually, years of boom and bust gave way to chronic decline. The herring became too scarce to be worth searching for.

Although the herring trade continued after the Second World War, fishermen began to realise that the bounty of the Irish Sea lay not near the surface where the herring swam, but on the bottom where there were scallops and Dublin Bay prawns. Soon Manx scallops were being exported to the Continent and the USA and eventually they constituted 90 per cent of the value of all 'fish' landed on the Isle of Man. In the 1950s, nine-year-old scallops the size of side plates were not uncommon, but gradually the specimens in the catch became smaller and fewer. The regulations dictated that they couldn't be landed until they were at least the size of a three-year-old. Subsequently, over 90 per cent of the catch was just three years old and few got beyond four.

The area that was being trawled every year in the Irish Sea was estimated to be two and a half times the total area of the sea bottom. As some areas are unfishable, on average the rest was being scraped three or more times a year. It was worse than that because the scallops were being caught not with trawls but with dredges, whose mesh bags were attached to a metal frame with large teeth up front to dig out the scallops. The seventy-five-centimetre-wide dredges were deployed in gangs of six either side of the boat to gouge the sea floor on a track as wide as a dual carriageway. The scars may have remained visible for months. The true extent of the damage they have done is unknown, but some delicate creatures that were once abundant are now rare – the fate of innocent bystanders. Over time in disturbed areas burrowing worms often increase in abundance and shellfish decline. But we don't know what sea bottom communities *should* look like, only how they have been left by the continual passing of fishing gear. Adjacent to the Port Erin Lab was an area of seabed out of bounds to commercial fishermen so that we could examine parts of the bottom after they had been dredged experimentally three or four times a year or not at all. In the undisturbed plots the fauna on the 'unploughed' bottom became more abundant and diverse. The scallop crop was soon five to fifteen times greater than in dredged areas, and twelve times more reproductive. It was sending out millions of larvae to seed the surrounding area and later, when the recruits grew, the fishermen's catches began to increase.

Andy Brand, one of my new colleagues at the Port Erin Lab, was also growing juvenile scallops in cages of netting, until they were large enough to be planted out into the sea to restock depleted areas.

Much is known about the effects of fishing gear on the commercial species it is designed to catch, but little attention is given to the fate of the by-catch, the non-targeted species that are thrown back. Brand was also studying this problem. Scallop dredges collect not only animals, but also rocks, and these tumble around inside the net as the dredge bounces over the sea floor. It is not surprising that the scallops it catches are often damaged. This isn't too serious as they are to be shucked before being sold. But what happens to the incidental captives such as whelks, starfish and sea urchins, or the undersized scallops that must be thrown back? There is little doubt they are frequently damaged, sometimes fatally so. When scientists took them into care in an aquarium, those with minor damage usually survived, those with major damage often did not, although whelks and starfish have remarkable powers of repair. Fishermen knew that starfish ate scallops so they chopped them up into bits before casting them overboard. Unfortunately the centre of the resilient starfish can rapidly regrow any arms that are missing. So fishermen were inadvertently creating several hungry starfish where there had been only one.

But the sea floor is no sanatorium. When dredge-damaged animals are discarded into the sea they fall to the bottom and leak 'juice'. Ever-alert predators home in on the scent and make short work of them.

Research divers went down to examine any casualties that were left on the sea floor after the dredge had passed. They were the same species as found in the by-catch but, because the dredge is so inefficient at catching these creatures, there were far more casualties left unseen on the sea bottom than discarded by the fishermen above. Some, such as the edible crab, were far more seriously damaged when clouted by a passing dredge than if rattled around in the bag.

Nowadays, fishermen seek their prey with electronic fish-finders and there is even sonar mounted on the net so that the skipper can guide it to enclose a school. Fish have as much chance of evading the net as a tree has of dodging the axe. And nets are getting much bigger: the gaping mouth of some trawls is wide enough to allow half a dozen jumbo jets to enter in formation without touching the sides, and giant purse seine nets could engulf St Paul's Cathedral. Some purse seiners also use spotter planes to locate the shoals. The Taiwanese fishing fleet alone deployed 14,400 kilometres of drift nets every night during the five-month-long squid season. These 'walls of death' are ten metres high and the mesh so invisible that even sharp-eyed dolphins, turtles and whales can't avoid entanglement.

Fish and shellfish are the most intensively hunted creatures in the world, yet ironically it is fishermen that are ultimately the endangered species.

Perhaps the most eerie thought is that even if all the fishermen became extinct, some of their gear would go on fishing. Every year in the North Pacific another 7,000 kilometres of drift net are lost, but they continue catching fish and entangling diving birds. One survey indicated that off Newfoundland over 100,000 birds and mammals were killed this way in only four years, and this estimate took no account of those that had been filched from the nets by predators or had simply decayed and fallen out.

Lost fish traps and lobster pots also go on fishing unattended. It is estimated that in this way $250 million of the commercial lobster resources in the USA are lost every year. The traps are also self-baiting, for the first animal to get caught and die lures in the second and so on, and on, and on. There are decades of deaths in an abandoned pot. We call it ghost fishing.

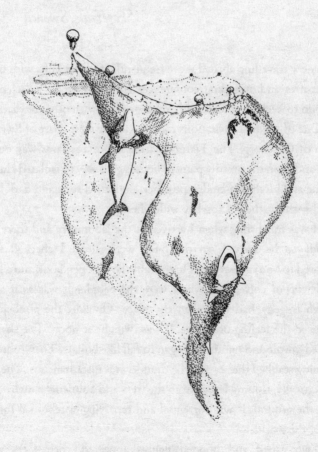

Darkness at Noon

Skagerrak, Sweden

I was now travelling abroad more frequently, forging links with other universities and attending conferences and workshops, as scientists must do to keep up to date and exchange ideas with other researchers.

One of the first invitations took me to the west coast of Sweden, north of Göteborg. The Tjärnö Marine Laboratory was way out of town and soon the prettily painted houses gave way to isolated clumps of pine and birch in a landscape of ice-smoothed rock and dark lakes edged with swathes of rosebay willowherb.

It was lunchtime when I arrived at the laboratory and there was nobody to be seen. Even the birds were silent. I checked what appeared to be a mess room. On the table lay an open book and a half-drunk cup of coffee, but where were the people? It was as if some sudden tragedy had overtaken the place. Outside, the atmosphere enhanced the feeling of unease. It was twilight at noon. The sky was bruised purple and raindrops began to fall like bullets. Then suddenly lightning stabbed the sea and the thunder was instantaneous. The rain fell vertically, ripping leaves from the trees and bouncing a metre high from the ground. It was torrential and terrifying, a rehearsal for the end of the world.

Two caped and hooded figures appeared, their faces mere

shadows. They leaned forward straining to haul a flatbed cart. On the cart was a dead seal. I was trapped in an Ingmar Bergman film.

Every day more dead seals were brought in, three corpses that first day, seven the next, twelve the day after. All along the coast were seals hauled out on the rocks, some lay alone, listless and wheezing, their eyes caked with a yellow discharge. Soon they would all be dead and no one knew why.

The cadavers were autopsied and tissues sent for examination. It would take months to discover they were being killed by a previously unknown virus only a billionth of the size of a seal. This 'seal distemper' attacked the immune system causing infected animals to fall victim to pneumonia and other diseases. It began here in the Skagerrak but spread rapidly, and by the end of the year 18,000 seals had died, half the entire population of harbour seals in the North Sea.

The outbreak is thought to have started when infected harp seals came south from the overfished Arctic, probably in search of food. The harbour seals had no resistance to what was, for them, a new disease and it ran through their crowded populations. The virus was related to those that give us measles and dogs distemper.

The death rate of harbour seals was much higher in those that were contaminated with pollutants such as organochlorines. Infected seals stopped feeding and lived off their reserves of fat, thus mobilising contaminants that had been stored away safely until then. This poisoning doubled their vulnerability to secondary infections. Ironically, in some areas it also allowed populations to recover more rapidly. In the Dutch Wadden Sea, where before the outbreak the seals had a low breeding rate, the demise of the most contaminated animals left a healthier and rapidly breeding population. However, fourteen years later seal distemper broke out again in the North Sea and, although by now many of the animals had some immunity, thousands died.

Similar viruses have since caused the death of dolphins, porpoises and whales, and have reduced the largest aggregation of the critically endangered Mediterranean monk seal to only 109 individuals. Other viruses, such as those causing brucellosis, have been isolated from seals, dolphins and porpoises. These are just a few of the wild animals that are falling prey to what have been called emerging diseases. Only a few years ago infections were hardly mentioned as a threat to endangered species, but emerging diseases that were innocuous in their place of origin proved to be devastating when they came into contact for the first time with populations of naive victims. In recent years, catastrophic epidemics have decimated once abundant populations of starfish, abalones, crows, vultures, antelopes and wildebeest, just as Aids and Ebola fever have done to human populations. And who knows what the future has in store?

In the same terrible summer that those first dead seals were being hauled from the Skaggerak at Tjärnö, it was not only blood that stained the water. In May and June, microscopic algae of just one species bloomed nearby and within four weeks they suffused the entire Skaggerak and Kattegat and were spreading up the coast of Norway to beyond Stavanger. These tiny cells reproduced not by making babies but by dividing in half and thus doubling in number. There is a story of a youth who so impressed the King that he was offered anything he desired as a reward. He asked if he might have a chessboard with a single grain of wheat on the first square, two on the second, then four, eight and so on. The King, who was not good at maths, said, 'Is that all?' and granted his wish, not realising he had just given away the entire harvest of his country. That is the power of doubling repeatedly. In the sea the concentration of algal cells reached 100 million per litre of sea water and each cell was not only consuming oxygen at night, so that fish and other creatures suffocated, it also manufactured a poison that damaged the gill membranes of fish who then suffered a lethal influx of chloride into the blood. Over five hundred tonnes of caged salmon and trout died, together with uncounted numbers of wild fish, and hordes of shellfish killed by dead toxin-laden cells sinking to the bottom. Insurance companies paid out $10 million in compensation to the fish farmers.

Curiously, this was the first massive bloom of this particular species ever recorded and although there were quite large blooms of microscopic nuisance algae in the Skagerrak every spring in the following years, it was not the culprit again until 1998. Harmful blooms caused by about seventy species of microscopic algae are commonplace around the world, and when they occur close to the surface, are easily seen by satellites in space. One massive bloom in 1995 stained the sea for over four hundred kilometres southwards from Santa Barbara in California to beyond the Mexican border. Often they turn the water brown, red or orange and are called 'red tides'. The Red Sea was so called because of the frequency of coloured blooms.

It was assumed that the leaching of nutrients such as nitrogen into the Skaggerak in the run-off from over-fertilised fields and planta-tions had stimulated that bloom, as they had in many other places. But this time it had little to do with our abuse of the sea. The alga had just taken advantage of favourable conditions. It happened to be there at the right time and its toxic secretions ensured that it suppressed the growth of rival species.

Many people claim that harmful blooms are on the increase, which may be true, but now that every fjord and sea loch has a fish farm, algae no longer bloom unnoticed in the desert sea. In the Skaggerak, fossils in the sediments revealed that blooms and mass deaths of clams probably happened periodically over the last 4,500 years, long before humans were significantly polluting the sea. Indeed, one of the plagues of Egypt was clearly a red tide: '. . . and all the waters that were in the river were turned to blood. And the fish that was in the river died; and the river stank, and the Egyptians could not drink of the water of the river; and there was blood throughout the land of Egypt.'

The Skaggerak bloom presented no risk to people, but some toxic blooms are very dangerous indeed. Many clams such as mussels and scallops filter microscopic creatures from the water and may ingest huge numbers of algal cells. Even the poisonous ones don't harm the clams, who merely incorporate the toxin into their tissues where it becomes concentrated.

The problem arises when a vertebrate consumes the clam, which is now ten times more toxic than strychnine. No matter how little clam meat is eaten, the result is rapid and sometimes fatal. There are at least five types of shellfish poisoning that arise in this way, and only diarrhoeic shellfish poisoning does not kill, it merely delivers severe nausea, vomiting and worse. The others result from different toxins and even when not fatal may paralyse the muscles or attack the nervous or circulatory systems, or induce amnesia, which can be permanent. Any vertebrate, including birds, may be affected if they consume contaminated shellfish. In 1991, pelicans and cormorants off Santa Cruz in California, acting as if drunk and swimming around in circles, were suffering from a form of shellfish poisoning. Old seawater samples indicated that this particular poison had been prevalent in twelve of the previous seventeen years. In 1961, just up the coast at Monterey, birds had flown around smashing into things and pecking at people, incidents that are now thought to have begun with an algal bloom and ended up in Hitchcock's film, *The Birds*.

Some types of shellfish poisoning were described for the first time as recently as 1976 and 1987, but paralytic shellfish poisoning has been known for centuries. Captain George Vancouver named 'Poison

Cove' in Canada after an ill-fated dinner of clams. Even then some of the local tribes knew that poisoned clams followed water blooms. They did not harvest shellfish when the sea shone at night, and many of the guilty algae are indeed luminescent. Other temperate countries shunned shellfish when there wasn't an R in the month, thus avoiding late spring and summer when blooms are most prevalent. In French Polynesia from 1960 to 1984, there were 24,000 reported cases of poisoning by algal toxins in fish, and it is estimated that at least three hundred people a year die from eating shellfish contaminated by toxic algae.

It is not because the bloom-forming alga *Pfiesteria* has an awkward name that it has been called the 'phantom of the ocean' and the 'cell from hell'. It's because it is a very scary plant. Although microscopic, it ambushes its prey. *Pfiesteria* lies hidden on the sea floor until fish aggregate nearby in large numbers. It detects their excreta and up it comes and starts producing toxins. At first these drug the fish so they become lethargic, until the poison reaches their nervous system and erodes the skin, creating bleeding lesions. Each of the tiny cells anchors on to the fish and inserts tiny hairs into its flesh and dissolves it alive. As soon as the fish die, the alga transforms itself into an amoeba-like creature that consumes the carcass. *Pfiesteria* is a master of morphing – it has no less than twenty-four different forms that appear during its life history. When replete, it cocoons itself back into the bottom sediment.

Although *Pfiesteria* is found worldwide and for much of the time is benign, in quiet warm estuaries where the water is enriched with nutrients it becomes virulent. In 1991, it killed more than a billion fish on the coast of North Carolina.

It even attacked the biologists who were studying it. They had tanks in which fish were being subjected to *Pfiesteria* attack, and didn't realise that the nerve toxin was rising into the air as an aerosol. Five researchers suffered severe but fortunately temporary amnesia. The team leader, JoAnn Burkholder, 'lost' eight days of her life. And, for a while afterwards, she couldn't read or hold a conversation or add 3 to 4. A colleague couldn't remember his name or find his way home, and even with reading lessons it took him three months to regain his ability to read and remember. Seven years later they had still not completely recovered their health.

Burkholder had to withstand a vociferous lobby from the main polluters of the estuaries who tried to discredit her research. Coastal developers and the local tourist industry were opponents because her findings were bad for business. Her funding began to dry up. It was not until a further outbreak, this time much nearer to Washington DC, that Congress and State legislatures acted to try to mitigate further outbreaks.

In my travels I would soon learn that, like coloured stains on the water, some governments were benign, others were dangerous.

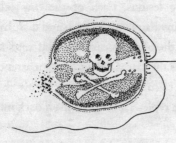

The Shipwrecked
Dentist

Hadramawt, Yemen

In my guise as a marine biologist, it is possible to visit places where few Westerners are welcome. The southern coast of Arabia is one of the least studied in the world and, when I had a student trying to assess some of the local resources of seaweed, I was invited to spend several weeks in South Yemen.

In poor, arid countries it is difficult to increase the productivity of the land, whereas the adjacent ocean is often highly fertile and nature provides the harvest. In such countries the population don't just eat fish; almost anything that grows in the sea is consumed. But, as everywhere else, over-harvesting jeopardises the crop. Third World countries are also always short of the means to earn foreign currency, but pharmaceutical companies are eager to screen thousands of tissue samples in the hope of finding just one species that contains a useful compound that can be marketed. Who would have thought that an insignificant sponge would have provided the most powerful anti-cancer drug we have?

South Yemen is now called the People's Democratic Republic of Yemen. 'Democratic' in a name is always a bad sign. At the terminal in Aden, I admired the arc of bullet holes decorating the wall.

Travel in the republic was strictly controlled. Until my permit arrived I was confined to Aden, a city that Vita Sackville-West had called 'precisely the most repulsive corner of the world'. The cemetery is full of those that had 'died prematurely on entering Aden'. This is the hottest coastline in the world and even a British brigadier had once admitted it was too warm 'to wear a tweed suit with any degree of comfort'.

A passing military parade had an unusual combination of bagpipes and Russian goosestep marching. I might never have known that it was celebrating the twentieth anniversary of the British withdrawal had it not been for hornets. Sitting in my room I noticed an insect emerge from a hole in the corner of the ceiling, then another. Within a minute there were thousands of them and I fled. The receptionist, who dealt with such things every day, whipped out two huge aerosol cans of insecticide and emptied them into my room.

I sat in the foyer and watched television documentaries recalling life under the British. The screen was filled with British soldiers beating locals over the head or firing into the air above angry crowds. The current news bulletins concentrated on the British firing at people in Northern Ireland, and anti-government riots elsewhere. Western democracies were ablaze.

I had arrived in the Yemen shortly after the assassination of the vice-president and three members of the politburo, and giant murals of the 'four valiant martyrs' were being painted on walls. Someone suggested that I might like to visit the museum that turned out to be largely devoted to 'the brutal years of British rule'. The guide made it clear that the Americans and British had also sponsored the recent assassinations. 'Here is the proof,' he said solemnly, pointing to the very weapons used to kill the martyrs. Why, in a country with four guns to every person, the assassin would use a rusty American hunting rifle and an antique English double-barrelled shotgun was a mystery. 'A Western imperialist plot,' the guide concluded and I nodded in agreement.

I longed to see the Indian Ocean rather than Aden's oily harbour and it was a relief when I was allowed to fly five hundred kilometres east to al-Mukalla in the Hadramawt. We followed the coast with the sandstone and lava desert on one side and the coral-rich sea on the other. Nature likes to juxtapose extremes such as desert and oases, and the ocean is the biggest oasis of all.

Immediately on arrival, soldiers confiscated the film from my camera and an emaciated dog threw up on my shoe. I was beginning to get a feel for the place.

What I had read about the locals in an old history book was not encouraging. Apparently there was not 'a race that exceeds them in evil and lack of goodness . . . blood of the slaughtered is everywhere . . . Hadramawt is called the Valley of Ill Fortune . . . All the women are witches. If a woman wishes to learn the most complete magic ever witnessed, she takes a human and cooks him until he dissolves and his flesh turns to gravy. When cold she drinks it all up, thus becoming pregnant.' What she gave birth to was more horrid that you can imagine and when mature it copulated with its mother and family values went from bad to worse.

Of course, that was in the old days. Now it was a different place entirely. South Yemen was fiercely tribal and tribally fierce. Kalashnikovs were openly traded in the market and were carried as casually as a rambler's walking stick. The Kalashnikov was one of the machines that forged today's world. For only a few dollars it gave anyone the means to fight back effectively.

I was assured that if the tribesmen weren't shooting at you, they were extremely hospitable to strangers. I was not convinced when I heard a traditional tribal chant:

> **We are the Awaleq,**
> **Born of bitterness.**

> We are the nails that go into the rock.
> We are the sparks of hell.
> He who defies us will be burned.

Al-Mukalla, the second city of the Republic, was strung along the shore and backed by precipitous cliffs only an alley's length away. Although breeze block was now making its aesthetic appeal evident, many of the houses were still built entirely of mud and chopped straw, which served for bricks, mortar and rendering. Those on the seafront were whitewashed, the rest were grey, enhancing the pink in the cliffs behind. Except in the walled garden beside the Sultan's palace, there wasn't a leaf to be seen. I assumed that truckloads of dust were brought in every night to recarpet the streets.

Two countries vied for the title of the poorest in the Arabian peninsula: the Yemen Arab Republic to the north and the People's Democratic Republic here in the south. Piped water would soon be brought forty kilometres to al-Mukalla, but sanitation was unknown. I noticed that pipes protruded over the street from the walls of every floor of the houses. These, I feared, were not merely to shed roof water.

When Freya Stark visited al-Mukalla she sympathised with a man whose child had died. 'It does not matter,' he replied. 'I have lots more.' The infant mortality rate was still a fifth of all live births. In 1962, there were only fifteen doctors in the whole of North Yemen and, outside Aden, there were none at all in the south. Illiteracy under the Marxist regime was still 86 per cent and there were almost no home-grown schoolteachers and even those only taught the Koran.

Most of the food had to be imported and little of anything was sold abroad. It was difficult to imagine that Yemen had once been the powerhouse of Arabia and a rich trading nation. The Kingdom of Saba (Sheba) flourished almost 2,000 years ago. It is now remembered only for the Queen who killed her husband on her wedding night and fascinated Solomon with her hairy legs. But in those days the world was awed by Saba's great six-hundred-metre-long dam at Marib that allowed the cultivation of 4,500 hectares of desert to become the 'garden of Arabia'. The Roman historian Siculus was impressed: 'Adjacent to this waterless and desolate land is another Arabia so much superior to it . . . that from the great profusion of foodstuffs and other material goods it has been named Arabia the fortunate.'

The dam lasted a thousand years. Although it was repaired in AD 450, a task requiring 20,000 men with 27,000 camels and donkeys, a hundred years later it collapsed and the desert returned.

The Queen of Sheba took spices as gifts to Solomon, for the Hadramawt was on the spice route from the east, and as Herodotus said, 'The whole country exhales more than earthly fragrance.' Later, the gifts of the Three Wise Men must have come from here, as this was the source of frankincense and myrrh. Gold was also mined in the region, but perhaps the Wise Men took not mere metal, but 'divine gold', the Sabaean name of yet another treasured spice. The people of the Hadramawt were the middlemen of the spice trade and no caravan passed through without their permission. Trade was brisk. Twenty-five tons of frankincense were sent to the Persian court of Darius. Emperor Nero was so distraught at the loss of his consort, he expended the whole of Arabia's annual production on her funeral pyre. Frankincense was processed in high-security factories. According to Pliny, 'Before the workers are allowed to leave the premises they have to take off all their clothes.' Now the trade and even the trees that bled the fragrant resin have gone.

From the Middle Ages until the nineteenth century, coffee (which comes from an Arabic word) sustained the economy. By the end of the eighteenth century, British ships brought the beans to

Britain directly from the Yemeni port of al-Makha, which they called Mocha. But now only 1 per cent of the land was still cultivated, its fertility irretrievably lost and Yemen's wealth long gone.

As a guest of the state, I stayed at Government House. I had envisaged a fading mansion from the last flush of empire, with decrepit four-poster beds and noisy plumbing. It turned out to be just a house owned by the government, a billet for council workers. My room was a trailer in the garden. The tin roof fluttered in the hot, heavy wind, but inside it was blessed with air conditioning. Every eight minutes the motor exploded into life or shuddered to a halt with a burst of simulated machine-gun fire. In a country where anonymous assassins might well kick down the door and spray the bed with bullets, I slept little.

Generous provision had been made for my food. A typical meal was a huge bowl of what I took to be mouse droppings and potatoes in hot, greasy water, smelling of spices and decomposition. It was followed by a dozen fried fish and a hill of grey rice. At intervals, the toothless cook emerged to inspect my progress and enquire whether I had enjoyed the mouse droppings. Although I tried my best, it made no impression. The cook worried about my health and, noticing his filthy fingernails, so did I.

With a temperature of forty degrees centigrade and 80 per cent humidity, what I really needed was a refreshing drink, but the coffee was viscous as treacle and twice as sweet. The workmen at the other table had large jugs of tea. Not knowing the word for tea, I pointed to a jug and nodded enthusiastically. The cook understood instantly and returned with a huge jug of coffee.

Shaving was difficult as there was no plug in the sink. There are no more than a dozen sink plugs in the entire developing world. Perhaps the others were shipped separately and lie mouldering in crates on sweaty docks, or have shrivelled to rubber raisins and been consumed by peckish stevedores.

My student Saeed smiled sparingly. Perhaps because his mouth contained more gold than an Inca burial. His lightweight suit was always immaculate, in contrast to my damp, clinging shirt and shapeless trousers. He never seemed to sweat, while I perspired like a guilty man in custody.

Saeed treated me with deference mixed with the gentle exasperation that might be shown towards a revered but slightly dotty relative. His most frequent phrase was 'There is a problem'. It was a struggle to get things done. The military regime was restrictive and easily roused, and the bureaucracy genially incompetent. We sat for hours in the armpit of sweaty rooms with a huge fan in the ceiling rotating so slowly it merely stirred the heat, as if poking the fire.

Eventually, we got to see the regional governor. In his presence, Saeed revealed that he was from the Uriah Heep school of deportment. His crumpled supplication worked. The governor offered us an official car.

To the driver's amazement, I put on the safety belt. It twanged into place with a cloud of dust, leaving a red diagonal stripe across my white T-shirt as if I were captain of a netball team. In shorts I stuck to the plastic seat covers and getting up was like peeling Sellotape.

Beneath a fur-covered dashboard the dust-encrusted rev counter was stuck at zero, the speedometer recorded only half the velocity experienced by the wheels, and the clock was two and a half hours slow. The new Russian-built coastal highway was wide and level with a reasonable surface, unlike the tarmac in Britain that turns to liquid on the first warm day of summer. It was covered with red wind-blown dirt that rose as an angry plume behind us as if we were on fire.

The tides unveiled the shores twice a day so we pursued them in the heat of noon and the soft warmth of midnight. It was arranged that the driver would take us to four shores one after the other. As low tide was at the same time on each shore and they were some distance apart, this meant that we made a rushed collection at the first site and a perfunctory one at the second. Although dashing on at reckless speed, we arrived just in time to see the third shore vanishing beneath the surf and the fourth enjoying high tide. This was clearly unsatisfactory, but it had 'been arranged' so we did it again the next day and the next . . .

All this hopping around in circles was reminiscent of the *nisnas* that, legend has it, once roamed just east of al-Mukalla. They were humanoid but with a face on their chest, and had only one eye, arm and leg. They were brought to Yemen by a local ruler called 'He of the Frights'.

When in the car we had to slow down as we passed military posts. The indolent militia, sweating in their helmets, just managed to look up. We nodded and smiled and they waved us on, like somnambulists swatting a fly. I never learned what they were guarding against whom. Their posts marked the division between one tract of dereliction and the next. Although I had an official permit to take photographs, every time I raised my camera Saeed said nervously, 'Oh no, Professor. This is a Military Zone.'

Saeed pointed out the passing highlights – a vehicle dump or a stagnant lagoon – with the words 'Beautiful, tourism', referring to some planned development. It never occurred to him that visitors could find beauty in the amber-and-chocolate desert or the wind-eaten bluffs eroded into rotted skulls and animal grotesques. One evening we parked on an escarpment beside a ruined fort left over from the 'days of fear'. As the sun died and the desert blushed in anticipation of the night, two women advanced over the nearest dune, carrying bundles on their heads. One was swathed in a robe of brilliant tangerine, the other wore electric blue. They were luminous and the effect was magical. No wonder Solomon's desire was to be here: 'I will get me to the mountain of myrrh, and to the hill of frankincense,' not to mention the Queen of Sheba.

In town, women were conspicuously invisible. Dressed head to ankle in black, they drifted through the dusty streets like shadows. Within their cowl they wore a dark veil that hugged the face, revealing its contours, but too dense to display even the brightest eyes. Only the hands and feet were exposed and I noticed that they wore golden sandals and their nails were painted scarlet. It was a tiny display, a delicate rebellion. After three weeks I became a devotee of women's feet.

Saeed's mother had done my laundry and I was invited to the house to thank her personally. It lay beyond the cemetery where lean and hungry dogs patrolled, waiting for the wind to blow away another layer of dust from the graves.

The draught through the latticed shutters was a relief from the oppressive night outside. I left my shoes in the hall and was ushered into a sitting room without furniture. The cream walls had a ginger-painted dado, as if the place had once been flooded with rusty water. The only softness came from lace curtains that festooned the walls, and a carpet on the floor.

Saeed's youngest brother arrived unexpectedly. In the absence of chairs, we were offered large cushions. They laughed when I sat on mine; they sat on the floor with the cushion against the wall. Saeed brought a tray of soft drinks and we chatted. His brother was training to become an architect. 'I will make big buildings that stay up,' he boasted. For a while I might almost have been one of the family, but I was not able to thank the mother, for the brother would not have approved.

On my last night we were to work the midnight tide again. Eventually my lift arrived, not the usual car, but a camouflaged

Land-Rover, and not the usual driver, but a surly soldier with a hip holster. Lacking English, he gestured for me to go with him. As there was no sign of Saeed, I stalled until the driver lost patience and bundled me aboard.

We raced along the highway, then without warning turned off on to the dark dunes. The driver's sinister appearance was enhanced by the dim green glow from the dashboard display. Why was he taking me into the desert? Was only one of us destined to return? I gripped my plastic ruler like a Yemeni dagger. I wouldn't go without a fight.

The vehicle swerved and foundered in the soft sand, the wheels spinning and digging shallow graves. My executioner took a small mattock and climbed out. I was stranded in a snake-infested wilderness with an assassin. Should I run for it into the night? But in which direction? While I was considering the options, he returned.

At the third attempt the engine coughed into life. The Land-Rover lurched forward, slewing across the face of a dune and pitching headlong into the night. Suddenly we were bowling along on level ground — we had hit the beach. In the headlight beams, hundreds of mole crabs out for their nightly constitutional scattered frantically like country pedestrians in city traffic.

We stopped at the end of the strand. On a rocky outcrop stood what I took to be a shipwrecked dentist flanked by two men, each with a trident and a net slung over his shoulder like a Roman gladiator. It was Saeed in the middle, his white lab coat was supposed to make him look more 'scientific'. His companions were fishermen friends of his. With the headlights extinguished, the moonlight turned the pale dunes into waves so that we seemed to be embraced by silvered oceans on both sides.

The tide pools swarmed with sea snakes. I had heard somewhere that anyone wading round rocky crevices should be aware of the danger. Well, I was aware all right, but it didn't help. Snakes on land in daylight are one thing, but those unseen in dark water are a different proposition.

The bite of some sea snakes is ten times more toxic than that of a cobra and there is no antiserum. They can kill a fish instantaneously, but a man might hardly feel the bite and, as the amount of toxin injected is small, it takes a while to have effect. Within an hour or so, the paralysis begins in the legs and slowly spreads up to the jaw, followed by nausea, vomiting, spasms, convulsions . . . he may not even know what killed him.

Fortunately, most sea snakes are docile and rarely bite unless provoked, except in the breeding season when they are 'vicious and aggressive'. Unfortunately, this *was* the breeding season, when they migrate inshore in large numbers. One observer reported that 'a solid mass composed of millions of snakes twisted thickly together and mixed with foam, was seen to form a line about ten feet wide and sixty miles long'.

I also read that 'most bites occur when wading in muddy shallow waters'. So we waded into the muddy shallows to complete our collections. Saeed heard a splash behind us and leapt in terror. We collided and tumbled into the water. He screamed and every sea snake in the neighbourhood swam for cover.

While we frolicked in the shallows, the fishermen had waded chest-deep to cast their nets in the creamy surf. They caught three lobsters and a dozen fish, which we barbecued on the beach. We sat until dawn watching the sparks rise and listening to the sea lapping like a thirsty dog.

Next day, as instructed, I turned up at Riyan airport at 5.30 a.m., two hours before my flight time. An hour later, the airport opened. Four hours later I boarded the rustiest aeroplane ever to risk rising into the air. According to the Prophet, *every* journey is a fragment of hell, but I had been reassured that at least the plane would not be crowded. 'They dare not fill it, sir, the runway is crumbling.'

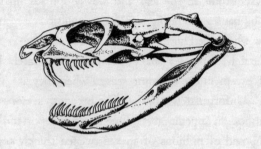

Piercing the Isthmus

Ismailia, Egypt

The Port Erin Laboratory had a tradition of link programmes with universities abroad. My colleagues had just completed a seven-year collaboration with the National Autonomous University of Mexico to develop prawn fisheries. Now they had secured a million-pound grant from the European Union to enhance marine research in Egypt. I flew out to cement relations with our Egyptian counterparts.

Foreign aid and income from tourists have kept Egypt's economy tottering along. My hotel was full of tourists 'doing' Cairo in a day. They rushed through the museum to view the glitter from Tutankhamun's tomb, albeit dimly through dusty glass. Then they were off to the Pyramids at Giza and, having thought they were in the middle of a desert, were surprised to find them in the suburbs. 'Would you believe,' I heard one say, 'all the beggars spoke English and the camel boys shouted "Tally ho!"'

The next day the tourists were off to see the 'real' Egypt of tombs and temples at Luxor and Karnak. It has been thus since 1885 when Thomas Cook first commandeered steamers to take his customers upriver into Egypt's deserted past. All the passengers were presented with a guidebook - *Notes for Travellers in Egypt* — a tome of 970 pages. Cook's tours ran like a military operation. Indeed, when General

Gordon had terminal trouble in Khartoum, it was Cook's fleet that ferried 18,000 troops to the rescue and, many said, would have made a better job of running the entire campaign.

Not being a tourist, I was going in the opposite direction, eastward. It was a dusty ride to Ismailia on a rickety bus whose only flypaper had been overwhelmed. We halted at a petrol station for a 'comfort stop', but the lavatories offered no comfort whatsoever. Unfortunately, a sign made it clear that FOREIGNERS ARE FORBIDDEN TO LEAVE THE ROAD.

Ismailia is a garden city, the 'emerald of Egypt'. Between the trees were open spaces where the grass was mowed by grazing sheep. The main danger was not traffic but the wiring of the street lights. On one road, bouquets of raw wires sprang from the lamp-posts. I trod carefully in case contact instantly extinguished the lights in both me and the business quarter. My fears were confirmed when the first rainfall for a year generated sparks that leapt across the road like horizontal lightning.

The town lies on the banks of Lake Timsah (Crocodile Lake), one of several lakes that form part of the Suez Canal. Here, ships gathered before entering the one-way canal in line. From the balcony of my flat I watched the enormous superstructure of a ship towering over the buildings and gliding past the end of the street.

The ancient Egyptians dug a canal from the Red Sea to the Nile to link it with the Mediterranean. It fell into disuse until rebuilt by the Persians, then the Greeks and Romans. The Persian canal was reputedly wide enough to be navigated by two triremes abreast — about thirty-five metres. On a surviving monument, King Darius proclaimed, 'With the power of Persia I conquered Egypt. I ordered this canal to be dug.' Its bed was still visible. The modern canal took ten years to excavate, cost twice the original estimate and opened in 1869. It 'pierced the isthmus' by stretching 192 kilometres to join two seas.

The journey time from Europe to India was slashed but the canal was not an instant success. In its first year only five hundred ships passed through, yielding a tiny profit. The Egyptian owners faced impending ruin. The chief shareholder, the Khedive Ismail, who gave his name to Ismailia, had accrued debts of £91 million, a fantastic sum for 1876. When he sold some of his assets, the British government acquired almost half the shares of the Suez Canal Company, which caused resentment in Egyptian nationalists for generations. Within ten years the income from the canal grew tenfold and was rising rapidly. But it was not only ships that were passing through.

Land is a barrier to the distribution of marine organisms. The faunas of the Caribbean and the Pacific Ocean are entirely distinct yet at one point they are separated by a strip of land only forty-eight kilometres wide. The Panama Canal connects the two seas, but is not the ideal corridor, for it is barred by frequent lock gates and for much of its length is fresh water. Marine organisms need salty water and therefore few have passed through.

The waters of the Suez Canal were also unwelcoming, but they weren't too fresh, they were too salty. The middle portion of the canal is not a canal at all, but a series of lakes. The Bitter Lakes are so-called because of their saline water. All tropical seas are saltier than temperate ones because so little fresh water runs into them and evaporation under the fierce sun is high; every year the surface of the Great Bitter Lake loses a layer of water two metres thick. To make matters worse, its bed is a natural deposit of rock salt. At the opening of the canal, this saline sump was almost five times more salty than the seas around Britain and was as effective as a dam for all but the most salt-tolerant marine animals, which became abundant. Within sixty years of the opening of the canal, the salty bottom of the Great Bitter Lake had dissolved away completely so the salinity of the water began to fall and marine organisms were able pass through the lakes.

There was, however, another barrier to cross. Following the annual

Nile flood a wall of fresh water formed at Port Said at the northern end of the canal and might have prevented some species from entering the Mediterranean. The building of the Aswan Dam curbed the annual flush of river water and that has ceased to be a problem.

The level of the Red Sea is well over a metre higher than that of the Mediterranean so the net drift of the current through the canal is from south to north. Also the Mediterranean tides are puny compared to those at the southern end of the canal where tidal currents can be too strong for divers to work in open water. These currents thrust the sea up into the lower reaches of the canal to the Bitter Lakes. Even on days when the currents are weak, the winds from the south encourage a northward drift of the water. In the last 130 years, over five hundred marine species have emigrated northward from the Red Sea, which has twice as many species as the relatively impoverished eastern Mediterranean. Only fifty species have migrated in the other direction.

Many of the newcomers have thrived in the Mediterranean. Half the catch of trawlers off the coast of Israel is migrants. The eastern Mediterranean is ideal for invasion. Its relatively sparse fauna and flora mean less competition for new arrivals and plenty of empty niches to be filled. The successful invaders are species that eat whatever is available, grow fast and devote much energy to reproduction.

The Suez Canal is now an open gate for dispersal. The Mediterranean receives five to ten new species a year from the Red Sea and the rate of immigration shows no sign of slowing down. It is called Lessepsian migration after Ferdinand de Lesseps who supervised the building of the canal.

Marine migrants were not the only invaders. The children of Israel were said to have passed through the mouth of the wadi where Ismailia lies. They returned, this time with tanks and overran the whole of the Sinai to the bank of the Canal. An Egyptian friend, who lived in Ismailia as a child, recalled having breakfast one morning in 1967 when the exploding shells shook their house and caused the tea in her cup to slop over on to the table. Her father rose and calmly said, 'I think we should leave.' They abandoned their house and did not return for six years.

Even when I was there in 1987, there was still a shell-pocked hotel and water tower as reminders of those dangerous days.

As part of a programme to develop eastern Egypt and the Sinai, the University of the Suez Canal had been established in a brand new campus at Ismailia. A member of staff invited me for supper. His musical doorbell played 'Baa Baa Black Sheep'. I had been warned not to eat beforehand because the spread could be enormous and hosts might be offended if a guest did not eat heartily. Unfortunately, on this occasion my host had settled for just tea and fruit cake and my stomach rumbled all evening.

Next day, my colleagues and I had an appointment with the president of the university. He was an absolute monarch in his kingdom and the only person that could make important decisions and most of the trivial ones. In the ante-room were four secretaries typing full tilt on the noisiest typewriters in the world. It was like being in a great factory, yet in two years of negotiations we had received no more than a couple of letters from the president's office.

The walls of the president's reception hall were lined with sofas on which to await his presence. The notion of an appointment at a stated time was considered a quaint Western idiosyncrasy. If there were later arrivals of greater importance, our appointment receded. An hour elapsed before the president came to our table. As our private discussion progressed, strangers from the adjacent sofa joined in.

Every Egyptian involved in the project had to be given an 'incentive'. This was understandable for the junior staff, who were so poorly paid. Unfortunately, our sponsors, the EU, had no concept of the institutional backhander. They were no doubt familiar with sweeteners delivered on a personal basis in private, but they could hardly be written into the proposal. So instead of 'incentives' the local collaborators received 'field allowances'. Even the president got a large field allowance, although he never left his office except in a limousine en route to a restaurant.

The university's public relations officer gave me a giant medal commemorating the founding of the university. As we went around the town he slipped medals into the palms of everyone from whom he wished a favour. Then I noticed the couple of notes beneath every medal that changed hands.

All he gave to me was a little book entitled *How to speak ENGLISH in 4 days. Without a teacher!* It covered indispensable English words including 'coution' (which I took to mean coitus mixed with caution), 'eye-water', 'nerf', and 'hand ballet', as well as phrases in common parlance such as 'Haw are you?' The list of professions omitted hawring, but included 'pat man' and 'traffic polish'.

A year later an Arab student that came to study with us at Port Erin went to register at the police station. Just as his passport was about to be handed to the sergeant, his alert supervisor retrieved a couple of tenners that had found their way between the pages.

Polyps and
Politics

Sharm el Sheik, Egypt

The four British staff hired for our project worked at Sharm el Sheikh, 350 kilometres south of Ismailia on the tip of the great shark's tooth of the Sinai that juts between the gulfs of Suez and Aqaba. It was their job to help establish a marine laboratory on the coast of the Red Sea and aid the local staff that would take it over. Fortunately, we had at last retrieved the two Land-Rovers impounded by Egyptian customs demanding exorbitant import duty.

It was agreed that we would depart for the coast at 10 a.m., but everyone arrived late and then went for breakfast. It was 12.35 when at last we headed south. Twenty minutes later we stopped for lunch.

There had been much discussion among the drivers as they secured the boat trailer to the tow bar of the Land-Rover. It was not a complete surprise when driving down the sloping highway to be overtaken by our boat. The trailer had come unhitched. We chased it down the highway until it swerved into the inside lane, hit the inevitable pile of bricks, slewed sideways and shed the boat. The outboard motor tumbled free and landed on its propeller.

The driver said, '*Malish*' (never mind). He was a companion to catastrophes. Only a couple of months before, he had driven off the edge of a raised road and somersaulted down the slope with seven students aboard.

The Sinai had been occupied by Israel for six years and although a peace treaty had been signed, all along the road were army checkpoints and UN observation posts. This was still nervous territory. The Sinai had been divided into three regions: 'Normal Egypt' – the ground recaptured from Israel; a demilitarised zone policed by the United Nations; and the rest, controlled by the Egyptian military. Even at Ismailia the university students had to have permits from the army to take samples from the Suez Canal.

The Sinai is an ochre-crusted desert stretching to distant pink mountains; a fabled place from where ancient Egypt took malachite and turquoise. It was inhabited by ten nomadic tribes for whom all non-Bedouin locals were strangers. Here, in the Bible's 'great and terrible wilderness', Moses received the commandments and hermits lived out their long days and nights. Almost 3,500 years later there had also been long nights for the soldiers that were crouched behind coils of razor wire in the gouged-out trenches and hidden bunkers marked by ventilation pipes. The international community had mobilised to remove thousands of landmines, but thousands more remained. In addition to these remnants of war, there were still occupied army bivouacs with light tanks parked alongside.

Sharm el Sheikh at that time was an untidy litter of back streets in search of a front street. Once past the final checkpoint on the outskirts it was patrolled not by the Egyptian Army, but by Multinational Observers. We stayed in the apartments at our marine laboratory in nearby Na'ama Bay. They were in a semicircular sweep on a rise behind the laboratory and one was sufficiently comfortable to be reserved for President Mubarak when he was in town.

The laboratories, originally built by the Israelis, looked good and the new dive unit was finished. We had an energetic team of young researchers in place and everything would have been wonderful had it not been for the bureaucracy. Unless forced into action, the local administrator did nothing in case he made a mistake and got into trouble. He was the keeper of the keys and refused to let anyone into the photocopier room in case they broke the machine. On weeklong trips to Ismailia, he took the keys to everything with him – just to be on the safe side.

The junior academics were justly proud of the little museum they had established and when a turtle came ashore it was earmarked as a prize specimen. 'You can't kill it,' I told them. 'Turtles are a protected species.'

'Yes, sir, but we do not have a turtle in our collection.'

'That doesn't matter,' I insisted. 'It's protected by law. We're not allowed to harm it.'

Clearly I didn't understand. 'But sir, we do not have one in our museum!'

The turtle was reprieved, at least until I was gone.

In front of the lab was a coral reef and when things got frustrating there was no better therapy than a dive among shoals of golden *Anthias* and scintillating chandeliers of glassy sweepers. Some of the most wonderful underwater landscapes in the world were within a few kilometres of Sharm el Sheikh. The reefs were stunning arrangements of coral plates and domes and ascending forks cemented with crusts of calcified algae enclosing a township of tunnels in which all manner of creatures resided. Some reefs formed walls of living coral plummeting vertically down from the surface for a hundred metres or more. Unlike many reefs elsewhere they were not periodically shattered by hurricanes and the arid coastlands rarely flushed down sufficient sediment to smother coral.

Many sites were so hauntingly beautiful that their names are known to divers all over the world: the 'Tower' at Sharm, the 'Wall' at Ras Mohammed, the 'Gardens' near Na'ama Bay, and the island of Tiran. Being close to deep water, many combined the near certainty of seeing ocean wanderers such as sharks and barracuda while enjoying the exuberance of inshore reef dwellers. A single glance could take in twenty species of fish and none had scrimped on coloration. The daily parade included grotesques such as the humphead wrasse whose expression was that of a man caught wearing a bowler at Royal Ascot. But most were undulating works of art, including unlimited print runs of Miros in motion, and butterfly fish that Bridget Riley had brought into line. The funky Picasso triggerfish was clearly a pastiche by Paul Klee, but there was no mistaking this season's flamboyant lionfish-look from Jean-Paul Gaultier.

The oddly oblong-shaped angel and butterfly fish might have been an aesthetic disaster had they not become canvases of colour. There are ten or so common species on these reefs, each with distinctive livery. The angelfish have names to match their robes: peacock, emperor, queen. Several are traversed by vertical or horizontal stripes, although one has them in irregular concentric circles like contours on a polychrome map. The stripes are usually blue or purple, sometimes outlined and accentuated in a different shade, often on a bright yellow background. Each species is as distinctive as a bar code, but just to make sure, they sport a variety of coloured eye masks and fins or tail in startling purple, orange or red.

This flamboyance is for show of course. Males often flaunt the brightest wardrobe and it is the showiest that wins the female. But why should a discerning female equate a dandy with someone who is fit and can bring home the aquatic bacon? Because she has read the colour code. The yellow, orange and red tints come from carotenoid pigments, and if bound to a protein, the greens, blues and purples can also be derived from them. Carotenoids are what make a carrot orange, and fish get them by eating algae or indirectly from crustaceans that have consumed algae and accumulated the pigments in their body. Thus the male's bright coloration is evidence of his excellent foraging ability – but there may be far more to it than that.

A coating of many colours might also advertise a healthy immune system. Carotenoids are known to boost the number of cells that fight off invading microbes. They also reduce inflammation and are antioxidants that detoxify the body of free radicals before they do any damage. Thus an animal rich in carotenoids is likely to be a healthy specimen. One with such a surfeit that he can fritter them away on decoration is indeed a fit fellow. On the other hand, a puny impostor simply cannot afford this luxury and must utilise what carotenoids he has for defence not ornamentation. It is known that in several species of birds stressed individuals have more insipid plumage than healthier ones, and that the bright red gape of a chick's mouth that shouts 'Feed me' or rather 'Feed *me* not my brother' is brighter and therefore more persuasive to its parents when it is given a dose of carotenoids.

Could the male be exhibiting a fatal flamboyance? How does he remain conspicuous to the female without being equally obvious to predators? Surely his gaudy display makes him vulnerable as well as desirable. Possibly not. The stunning colours revealed by the flash of an underwater camera are misleading. Even in the shallows they are muted, and deeper down, where much of the red light is absorbed by the filtering water above, the red patches look black. The bars and stripes so conspicuous close up in bright light may blend into the background or commingle with the dancing ripples of light thrown down from the surface.

In any case, what matters is not *our* impression, but what potential mates or predators perceive. Surprisingly perhaps, at least some species of fish may have better colour discrimination than humans, although it might be different. Fish in general seem to be better at perceiving blue light than we are and many have special sensors for ultraviolet, to which we are blind. They may also change how they themselves are perceived by a fishy viewer simply by parading against a plain background when displaying to a potential mate, or hovering in front of the sunlit and shadow backdrop of the reef when hoping to remain unseen. It is also likely that some colours and patterns are apparent to fish if close to but not from a distance, so that the displaying male might show off his wares only at the appropriate range.

Juvenile angelfish frequently have different colours or markings from the mature adults, for the life-and-death lies of camouflage take precedence before the need to reproduce raises its risky head. Many species rapidly change into a more disruptive patterning when nervous or disturbed. Shoals of fish may change colour in synchrony or out of phase with one another and either way this may make it difficult for a predator to select a target. We have much to learn about the complicated business of disguise and display, but now at last we are trying to look at it through the eyes of the fish.

Organised leisure diving trips first came to the southern Sinai in the late 1960s, mostly on Israeli dive boats, but the war interrupted them and they stopped altogether when the Sinai was handed back to Egypt. The Egyptian State Tourist Ministry was keen to revitalise diving holidays, but the military that ruled the Sinai were obstructive. When I first visited, the once sleepy fishing village of Sharm el Sheikh was just beginning to grow into a centre for diving. In one of the hotels I chatted to a group who worked at local dive centres. They were from all over the globe: Australia, England, Germany, New Zealand, and had come all this way to talk of nothing but diving. There was a TV in the bar showing continuous videotapes of underwater adventures. One followed the antics of an expedition that included a former Miss Universe who was surely too pneumatic to sink. She screamed when confronted by a great white shark . . . or a jellyfish, or a worm. I suppose that aggregations of golfers and skiers spend their evenings watching videos of the hundred best putts or pistes.

I asked the people who made their living from diving if they worried about the effects of so many visitors. Those who knew the reefs best were the most dismayed. One said that the 'Temple' reef at Ras Um Sid was 'only a shadow of its original glory in the seventies' and a couple that had run a diving school here for twelve years claimed that a popular site near Na'ama Bay had 'been destroyed'. As the number of visitors increased it would get worse. In only four years from 1986 the number of hotels increased from two to twelve, plus four residential dive clubs and two large campsites. Surveys showed that at one reef the number of broken corals doubled in a single year when a permanent campsite opened nearby. Over twenty kilometres of this coastline had been designated for tourist development. By 1995 there would be forty international hotels, twenty-seven dive centres and fleets of dive boats. Now everywhere you look the seafront has been developed.

Most visitors came to dive or, worse, to learn to dive, and much of the underwater activity was concentrated in only ten or twelve locations. Even sixteen years ago, at the three most popular sites there were 1,455 dives in a single week in May, and that figure only included those organised by the dive centres, not the independent divers. Now 28,000 divers a year visit the popular sites. When I first dived at Ras Mohammed, I don't recall seeing another person in the water apart from my dive buddy.

A three-year survey in the late 1980s showed that there was on average ten times more broken coral and twice as much dead coral at the popular dive sites than at adjacent, less frequented ones. The reef top was also damaged where snorkellers walked across it to get to the sea.

Coral seems hard and tough, but is really a fragile structure covered by a living nap of feeding polyps. One tyro diver standing on a boss or simply grabbing a branch to steady himself in the current inflicts damage. Divers hypnotised by whirling rainbows of fish don't even notice that they have bumped into the reef. And who knows what those long insensitive fins are doing at the other end of the body from their eyes?

The explorer's (and the author's) dilemma is that, just as a rose petal is bruised by handling, entire living communities are vulnerable to the touch of humans. If we celebrate the beauty of a reef or a lagoon, we expose it to the dangers of excessive admiration. Nature thrives best on neglect.

When neglect is no longer an option, protection must come into play. The most famous sites at Ras Mohammed were given the protected status of a national park in 1989. Later, perhaps too late, the protected area was extended. The authorities banned spearfishing and souvenir collecting, and fined or imprisoned those caught taking large amounts of coral. Promiscuous anchoring of boats was forbidden and a hundred permanent moorings were installed over the sites. Walkways were constructed over the reef flat to reduce further damage from trampling. In other areas nearby new regulations limited coastal development, but allowed traditional fishing to continue. The Israeli Nature Reserves Authority now controls five additional areas further up the Gulf of Aqaba.

Overuse is a problem wherever there are coral reefs. An article that appeared in the *National Geographic* magazine in 1962 extolling the wonders of a coral reef state park in Florida preceded an invasion by tourists that never let up. There were 3,000 visitors a day at some sites; now one reef alone must suffer 150,000 dives a year. A later article in the same magazine chronicled the almost total destruction of the reefs by 1989 from pollution, careless boat use and diver damage. The article is full of heartbreaking before and after photographs. The damage of a day may take years to repair; the damage of every day is irredeemable. Excessive nutrient run-off from land into the Caribbean is compounding the problem by encouraging bacterial infections that have destroyed perhaps 80 per cent of the corals in the last twenty years.

Ships added incursion to injury; on the southern coast of the Sinai there is scarcely a reef that has not been gouged by one, or several. Local reefs seem to have a magnetic attraction for vessels. In a four-year period in the late 1980s three large ships rammed the reefs around Sharm el Sheikh. In 1988, the leaking *Lania* coated fifty kilometres of reef with oil. It took four days for the authorities to decide to spray the oil – with the wrong dispersant. A year later the *Safir* was deliberately grounded at Ras Nasrani after colliding with a reef in the Straits of Tiran. I was told that the master blamed the reef. Initially, it pulverised two hundred square metres of the reef flat. A week later the reef collapsed beneath its weight, sending huge blocks of coral avalanching down the reef slope destroying everything in their path. The ship sank into the depths, leaving behind a trail of toilet bowls from a smashed container. It also shrouded the reef with phosphate-rich sediment from its cargo to compound the damage and slow the recovery.

Cunard's top luxury liner the *Royal Viking Sun* veered into a reef off Tiran Island in 1996 and as she struggled to get off, the hull ground across the reef and bulldozed tonnes of coral down the reef face. More than 2,000 square metres of coral were damaged. The site was described as 'a graveyard'. Jackson Reef, not far from Sharm el Sheikh, had become far more conspicuous since the *Louilla* was thrown up by a storm in 1981 to perch upright on top of the reef, but that did not prevent several other vessels blundering ashore there and leaving a scatter of steel plates and debris. In 1999, a 21,000-tonne German freighter sliced the *Louilla* in half, leaving its unstable bow section to scrape the reef anew. The Egyptian authorities impounded both the German freighter and the *Royal Viking Sun* until they received compensation for loss of tourist revenue, $23.8 million in the case of the *Royal Viking Sun*, but only time will repair the reef. Providing, of course, no other ships decide to rush ashore to renew the damage.

As for the marine laboratory at Na'ama Bay, it sank as surely as a torpedoed ship. After my colleagues had spent years making it a research place to be proud of, the sheikh of Abu Dhabi made an offer for the site that the Ministry of Tourism couldn't refuse. Now a luxury hotel stands there and the remains of the reef that fronted the laboratory are not worth visiting. The Ministry promised to build a replacement laboratory closer to Sharm and eventually they did, but the site was nowhere near as suitable for marine research and the buildings are, I'm told, used mostly by the business community.

Dynamite and
Cyanide

Panay, Philippines

It is not only the reefs of the Red Sea that are endangered. Environmental degradation and declining fish stocks are problems all over the world. Perhaps half the research projects at my laboratory were conducted abroad and I spent time in Pakistan, Chile and the Philippines helping to set up research collaborations. Our efforts in Mexico and Egypt had been well funded, but were shackled by stultifying bureaucracy and the lack of a work ethic in many local scientists. As one of my colleagues put it: 'We worked, they watched.' In the Philippines, however, our collaborators were wonderfully eager and energetic, but we never secured sufficient funding to fully exploit their potential.

The jet eased out of an expanse of cloudless blue. Below me lay a country of 7,000 islands, eighty languages and dialects, and a million possibilities, most of them unrealised. We came in over the outstretched arm of Bataan that had grasped Manila Bay from the sea. The bay was so large that all the ships in the world could anchor here without making it seem cluttered. The shallows *were* cluttered with long fences erected to trap fish. They were shaped like flights of giant arrows indicating the direction of the currents. The sea mattered here.

Very little of it was left unused. If fish weren't being hauled out, sewage and rubble were being dumped in, often all three at the same time.

I knew the rules. Never pick up a taxi from the airport except through the official tending the rank. If the driver fails to drop the flag on the meter, refuse to go. If a local hops in as you are about to depart, get out. I travelled with a large diving knife in my bag.

Traffic in Manila was indistinguishable from parking with the engine running. It was the noisiest inactivity imaginable. The traffic lights were not working because there was one of the daily power cuts. The bus in front bore the sign, NO CLINGING AROUND, yet its exterior was festooned with non-paying passengers. It was quicker to walk.

To protect the traveller my driver had crucifixes and religious pictures dangling down the windscreen in such profusion they obscured his view. Drivers greeted you with 'Hi, Joe!' They assumed any lone, non-Japanese foreigner was an American and he was here for sex.

'Wanna girl?'

'No thank you.'

'Wanna boy?'

'No.'

'Wanna *little* girl?'

If you let him, he would rapidly run down the menu ending with the depravity of the day. But he didn't take offence if you really did want to go to your hotel. The choices might include transvestite youths called biniboys, or binigirls disguised as boys and, for all I know, bi-biniboys masquerading as girls pretending to be boys. Nothing is ever what it seems. Even at the Debonair barber's shop for a few extra pesos you could have your groin caressed with a vibrator while your hair was being cut.

There were over 100,000 prostitutes in Manila alone and that did not include males or children. Even veteran prostitutes were usually below the age of consent. Prostitution was of course illegal, but a notorious hotel for paedophiles was owned by the Minister of Tourism. The redder-than-red-light district was flanked by two streets, Pilar and Mabini. Ironically, Apolinario Mabini drafted the original constitution that gave women the vote.

The Philippines is the most Americanised country in the world. Just around the corner from Shakey's Pizza Parlor was Dunkin' Donuts and down a back street was a tiny shed, the Dew Drop Eatery - *Deadly weapons not allowed inside*. A copulation of the two cultures gave birth to banana ketchup. I watched the Academy Awards' ceremony on TV. It was indistinguishable from the American Oscars except all the statuettes went to Filipino actors and movies. The only trace of England I heard was from a truck whose reversing beeper played 'London Bridge is Falling Down'.

I am intrigued by the enthusiasm with which nations have come to embrace the culture of those who once oppressed them. In the Philippines it is a forgotten tale. The United States had missed out when the great empires were being amassed by Europe, until it wrested the Philippines from Spain in 1898. Commodore Dewey cabled home: 'Have captured the Philippines. What shall we do with them?' It was a question that would perplex the Americans for the next forty-eight years.

The Filipinos rebelled and the US Army became mired in a guerrilla war so they got tough. Even to utter the word 'independence' was punishable by imprisonment. General 'Howling Jake' Smith issued orders to 'Kill and burn, the more you kill and the more you burn the more you please me'. His men were told to take no prisoners and to slaughter 'everything over ten'. The Secretary of the War Department reported that the campaign had 'been conducted with marked humanity and magnanimity'.

Outside the American concentration camps was a 'zone of death' in which all property was torched. A congressman boasted that 'They never rebel in Luzon any more because there isn't anybody left to rebel'.

The United Stated eventually triumphed and embarked on what was called 'benign imperialism'. In 1945, the Republic of the Philippines was born and Filipinos became progressively urbanised and Americanised. But in this land of contrasts a few natives never got the message. A tribe inhabiting the hills only three hundred kilometres north of Manila kept up old traditions. While I was there the local newspaper reported they had beheaded two intruders. Only a few years before, when this was a more common occurrence, picnickers were alerted with the headline: IT'S HEADHUNTING TIME AGAIN. President Ferdinand Marcos, having seen how well other presidents looked on Mount Rushmore, had his giant likeness carved into an escarpment. But when the time comes for heads to roll, even the biggest head is vulnerable.

When in turbulent countries, I was usually able to avert my eyes and concentrate on the struggle for existence in a tide pool. But over the course of my visits the Philippines underwent a momentous upheaval that ended twenty-one years of dictatorship. It was impossible not to become embroiled in history.

Marcos had swept to power on a wave of expectation in 1965. He was charismatic and had Imelda, a beauty queen, for a wife. By his own admission he was the most decorated soldier in World War II and had a chestful of unearned medals to prove it. In the presidential palace you could see a huge painting of him emerging bare-chested from the jungle brandishing a *boloe* knife. Alongside was a portrait of Imelda emerging like Venus from the waves on a scallop shell. Perhaps they reminded them of how beautiful she once was and how courageous he might have been.

Marcos was going to put things right. 'We are in crisis,' he declared. 'Only by severe self-denial will there be hope.' What he didn't make clear was who would be denied. One of the certainties of Filipino life is that the rich get richer and the poor more numerous.

A president may only serve two terms of four years, so Marcos declared martial law, suspended the constitution and stayed in power for another fourteen years. Martial law was so popular with the Americans that they doubled their aid.

Political murders had been commonplace for years, but the killing of the opposition leader Benigo Aquino in broad daylight, the moment he stepped on to home soil, was so brazen it showed how contemptuous of public opinion the regime had become.

Marcos called a surprise election to reassure the Americans he was still the people's choice. Dictators only hold elections they are certain to win and Marcos knew every trick required to fix the result. Elections here had always been decided by 'guns, goons and gold'.

But Aquino's wife, Corazon (Cory), stood against Marcos and the victims of the regime identified with the brave widow. Marcos mocked her for being a housewife and she admitted that she was indeed inexperienced at lying, cheating, stealing and assassinating opponents.

When Marcos was declared the winner, Cardinal Sin, the Bishop of the Philippines, condemned the election as a sham and called for civil disobedience. The people took to the streets. Marcos was incandescent: 'I will set the tanks on them . . . I am not bluffing.' But the protesters lay down in front of the tanks and the soldiers refused to fire on civilians. Two days later Marcos and Imelda fled into exile. The Philippines, whose major export was women who went abroad as nurses, domestics, prostitutes and mail-order brides, inaugurated Cory Aquino as its first democratic president.

*

I flew south to Iloilo on the island of Panay. The local biologists at the University of the Philippines in the Visayas were delightful and, as in many developing countries, all but two of them were female because the salary of a university lecturer was insufficient for the main breadwinner of the family. Filipinas have a desire for diplomas and several of the staff came across to the Isle of Man to study for masters' degrees. One gained a PhD in cell biology at Liverpool University. Back in Iloilo there were no facilities to do cell biology and she was researching the possibility of cultivating worms for fish food.

Much energy was expended in planning. My suggestions were always enthusiastically received: 'We need a master plan for the programme,' they would say. The dean set up a task group, a feasibility study and a steering committee. The plan was sound and the documents were immaculately drafted, but everyone knew that it was unlikely to be funded. Filipinos are so accustomed to disappointment that they take it in their stride.

The students were less willing to accept the way things were. All over the campus there were posters:

> YOU ARE ENTERING THE UNIVERSITY
> OF THE PHILIPPINES
> A NUCLEAR WEAPONS-FREE ZONE

A respected American journalist had stated that the bases were there to act as magnets for enemy attacks so that the nuclear strikes were dispersed, thereby reducing the threat to the mainland United States. Young Filipinos had no desire to be targets.

Perhaps the expectations for Cory Aquino were too high and the problems insurmountable. Almost all officials were incurably corrupt. The police, I was told were 'the nicest people money can buy'. Even firemen attending a blazing hotel demanded a tip before attaching the hoses to a hydrant.

Over 70 per cent of Filipinos lived in abject poverty and still do. Twenty per cent died at birth and three million of those unlucky enough to survive became squatters. In Manila alone, 60,000 lived by scavenging on garbage dumps. Marcos's government had responded by erecting giant billboards to hide the shantytowns from view.

If the poor were desperate, the rich were nervous. There were twenty kidnappings a month for ransom. Seven of the Yellow Pages were devoted to guns and ammunition and twelve to security agencies 'specialising in anti-kidnapping' and 'VIP protection'. Beneath a picture of a wife and children cowering before a gun was the caption 'Prevent this happening to your family'. In a country where nothing worked, guns were always well maintained.

Most shops were tended by an armed guard; a large store might have a dozen. The house where I was staying sat inside a high wall with a uniformed guard on the steel gate day and night. He only came up to my shoulder, but with his pump-action shotgun he seemed much taller.

In her first year and a half in office, Aquino survived five coups. Even the now dead Marcos was not buried but refrigerated so that perhaps when medicine improved he could make a comeback. But the millions he had stolen were gone for ever.

In such parlous times the University of the Philippines had just built a brand new campus at Miag-ao, forty-five kilometres from Iloilo. It had been funded by a $1.9 million loan from the World Bank. It was to be a centre of excellence for fisheries research and the College of Fisheries was moved there from Manila. Unfortunately, there was no telephone line, nor a pier or sheltered anchorage for the research ship.

Understandably, poor countries put their money into applied

research that will benefit the economy. So although the university library was a sanatorium for worn-out books, the Fisheries College had an excellent library. While the biologists' research equipment consisted of a microscope, a chemical balance, an oven and a huge heated box for drying pressed plants, the fisheries laboratories were well equipped. The head of the Fisheries College confessed they lacked basic knowledge of the ecology of commercial fish and the biologists might be able to help. That was not to say that he would let them have access to his equipment or library. With funds from the British Council we bought a substantial number of new books and sent them from Britain, but they were impounded by customs. As the university could not pay the 'duty' they were never seen again.

The biologists worried less about the lost books than about me being alone in the house. They fixed up a companion to keep me company. Being so sociable they had no concept of solitude – to be alone was to be lonely.

One night we all went out to dinner at a karaoke bar. The chancellor of the university had by far the best voice and entertained us with 'Puppy Love' and 'Strangers in the Night'. There was no rock scene in the Philippines; even the young were hooked on show tunes. Surely teenage rebellion is rarely fomented by singing 'My Way'.

Nonetheless, rebellion there was. The New People's Army, the military wing of the banned Communist Party was in the hills. Driving to the new campus at Miag-ao to attend a student reunion, I noticed that every bridge had an NPA slogan painted on it. Many of the NPA were middle class and I met several of them at the barbecue. They had left their weapons behind and come down to chat to old friends at their Alma Mater. Typical of them was Nestor, a thoughtful committed young man who believed Aquino was too cosy with Marcos's old compadres – her Minister of Defence was Marcos's cousin.

The government had declared a National Marine Reserve on Taklong Island just off the coast and gifted it to the university. Getting there was a short voyage in an intermittently motorised outrigger canoe with no freeboard, followed by a long journey in a jeepney with braces of live chickens dangling from the door handles outside. The chrome-encrusted vehicle was called 'Lord Have Mercy' and now I knew why. The roads were rocky ribbons of choking dust. To permit breathing, everyone wore a bandanna across mouth and nose like a bandit.

The university had a laboratory on Taklong Island, a small bungalow with a generator for power, and bunks so we could stay over the weekend. There were no other facilities as yet.

The first thing I did was to go snorkelling in the bay and was astonished at what I saw. Instead of a colourful coral reef as bright and sparkling as the beach and palms, here below in the green haze there was an air of unremitting melancholy. It was a field of boulders, some decorated with soft corals resembling folds of pallid corduroy or clusters of grey hands raised in prayer.

This was the result of what is called fishing in the Philippines, not fishing with a line or net, but with dynamite. I chatted to a local called Boboy. I assumed he was a fisherman – he had two fingers missing. Boboy made his own dynamite from a mixture of fertiliser and fuel oil and poured it into beer bottles. He had lost his fingers cutting through a detonator to attach the fuse. When fishing, timing is everything. Once the fuse is lit, the trick is to hang on long enough for the bomb to explode just after you lob it overboard, before the fish can scatter. But when a local using this technique lost half his head, Boboy decided to first 'seed' the water with bait to attract fish and use a longer fuse so that even if the fish fled when the bomb splashed in, they had time to return before it went off. His bomb caused a mighty 'whump' and an eruption of white water at the surface. Boboy slipped into the water and I followed. Almost everything within a fifty-metre radius had been killed. We swam through a shower of silver fish, some still twitching. Many floated to the surface, but had he not dived, half the catch would have been lost. He gathered up all he could find, no matter how small. Big specimens were scarce now.

The effects of such explosions were described by the naturalist William Beebe, who never hesitated to blow up a pristine reef to collect specimens: 'I returned to a scene of desolation and found a shambles of coral, all covered thickly with grey dust . . . One great piece of coral weighing hundreds of pounds had been blown over on its side.' He clambered over the wreckage and realised that 'not a worm was injured. What poppies in Flanders took months to do, hundreds of worm blossoms had accomplished in an hour.'

Clearly, creatures lacking a gas-filled swim bladder are not so badly affected as fish. A student volunteer with air-filled lungs who was surveying a reef not far from here had been accidentally disabled by a bomb.

As human beings well know, even the survivors of bombings may be rendered homeless. If the architecture of the reef is destroyed so that it is no longer a labyrinth of tunnels and holes, it cannot house a large community. Life moves elsewhere in search of the crannies it needs to hide away from predators, and soft corals and seaweeds take over.

Dynamite fishing is of course illegal, as is what the locals call kayakas fishing, a more systematic method to pulverise a reef. I saw kayakas fishing not far from the campus at Miag-ao, almost within sight of the Fisheries College. There were dozens of skin divers moving in line a cross the bottom, each thumping the sea floor with a large weight to scare fish and shepherd them into a huge chamber net. The divers follow them right inside the net and if, as sometimes happens, they don't find the small escape opening within the duration of a single breath, they are as doomed as the fish.

The reef poisoners are also at risk. Their dive masks are supplied with air from generators in the boats above. They squirt cyanide or bleach from a washing-up bottle into the reef to anaesthetise fish, then break open the coral with a crowbar to get at them. It is estimated that there are over 6,000 divers inserting poison into 33 million coral heads annually. The animals are destined for the US and Europe and must be alive. Around 20 million tropical fish, 12 million stony corals and 10 million starfish, snails and crabs are sold annually for the aquarium trade, although well over half probably die as a consequence of capture or transportation. One in ten of the divers also end up disabled or dead thanks to the bends.

These are not the only threats to the reefs. The huge façade of the two-hundred-year-old fortress church at Miag-ao bears a wonderful bas-relief of the Spanish defeating the Moors at the Battle of Tetuan, probably as a warning to the locals. Most houses hereabouts are constructed of bamboo and palm leaves, but the church is built of coral, the only easily quarried rock that is available. Reefs are excavated with crowbars and taken away on sledges hauled by buffalo, but James Hamilton-Paterson described an excavator being used to take large bites out of the reef and dumping them into a truck that when heavily laden crunched back across the reef.

The problem is not confined to the Philippines. As tourism grew in the Maldives the demand for building materials was met by entirely denuding a couple of large reefs. The top half-metre of the reef flat was removed from atolls, 80 per cent of whose land mass was only a metre above sea level.

The waters of the Philippines and Indonesia have 20 per cent of the world's coral reefs and were so rich that a single reef here once housed more species than found in the entire Caribbean. Now less than 5 per cent of Philippine reefs are in pristine condition. On an island only a few miles along the coast from Taklong we found beautiful reef. The island had until recently been a military zone and was therefore a place that dynamiters and cyanide fishermen dared not visit.

It was in the Philippines that I first saw odd bone-white corals standing out against the colourful backdrop of a reef. Most corals are a liaison between polyps and tiny algal cells that provide them with carbohydrates, making them in effect photosynthetic animals. If under stress, corals evict the pigmented algae and therefore become blanched. This happens to individual colonies from time to time, but biologists began to notice widespread bleaching over entire reefs. According to the Worldwide Fund for Nature, such events had been recorded only three times anywhere in the world in the hundred years preceding 1979, but there were sixty reports from 1980 to 1993. It was taken as a harbinger of disaster, the final straw. The reports and predictions are frightening: '27% of the Earth's coral reefs are dead and 70% will be gone within 50 years.'

The main culprit seems to be a warming of the sea, whether sporadic, or perhaps a permanent consequence of global climate change. A reef biologist in the United States is suing his government on the basis that its policies encouraged the burning of fossil fuels leading to global warming and putting his career in jeopardy. If it is too warm and sunny, the algae go into photosynthetic overdrive and probably flood the coral issue with dangerous free radicals, so that it jettisons the algae. When the temperature drops, algae can be readopted and all is well, but if the warming is sustained, the coral may starve.

It is easier, however, to report coral cemeteries than to monitor whether the reefs are subsequently recovering and, if so, at what speed and to what extent. Thus reports of their death may be premature. Half of the severely damaged reefs seem to be recovering although the rates of recolonisation are very variable. Most of the Great Barrier

Reef, the largest living structure on the planet, is for the time being still in reasonable shape. This is not to dismiss concern over the dramatic decline of reefs, but to warn against overstating the problem. If dire predictions fail to come true, both governments and the public will suspect that the prophets are either far less expert than they pretend or have deliberately hyped the spectre of environmental disaster as a means to attract research funds.

In the late Permian era, a catastrophe, the cause of which remains a mystery, eliminated 96 per cent of all the animal species in the oceans. The corals, though decimated, survived and proliferated over the subsequent 245 million years in spite of four further catastrophes and severe climatic fluctuations. So they have shown remarkable resilience.

Global warming may heat up some seas whatever we now do, and it will undoubtedly be yet another major setback for the corals. But much can be achieved by seeding damaged reefs with small pieces of coral to speed up their recovery and by curbing pollution, overfishing and destructive fishing methods. The problem is how to do it.

Outright bans don't work in developing countries, for their coastlines cannot be adequately policed. They merely provide a means by which those who are supposed to enforce the ban can extract bribes from the peasantry. Worse still, they make an enemy rather than an ally of the locals. Paupers cannot be expected to obey rules that would prevent them from feeding their families, and what gives us the right to tell them to do so?

The only effective guardians of a reef are those whose future depends on it. Providing training and fishing gear would render poison and dynamite less tempting, but the locals also need a regulated market for their catch. Collecting for the aquarium trade is very profitable, but the fishermen receive a pittance, so they overfish to make ends meet. Controlling the trade by offering a fair price for the animals would reduce the exploitation of both the reef and the fishermen. In the Philippines, the Worldwide Fund for Nature is proposing some of these ideas to address the problems of coastal overfishing, and on the island of Cebu an alliance of local communities, divers and government is establishing protected areas.

It seems obvious that this is a social problem rather than just an ecological one. Hundreds of millions of the poorest people on Earth depend on wildlife for their existence. They must be given incentives to allow them to exploit and nurture both the animals and the environment on which they depend.

Part of the problem in the Philippines is that there are too many fishermen. Many of the dynamite fishermen had been farmers, but President Aquino failed to solve the age-old problem of peasants who owned no land and worked only to increase the wealth of landowners. To make things worse, logging had increased erosion and rendered marginal agricultural land unproductive, so recently the numbers of coastal fishermen doubled. Most of the dynamiters had no fishing skills; they were farmers who wanted to farm, if only someone would give them land.

Aquino had come to power with the cry 'Land to the tiller'. Yet at the time of my last visit to the Philippines she refused to meet a delegation of peasant farmers and set the army on them. Nineteen were killed and many more wounded.

At a dinner in the British Embassy I sat next to one of the politicians who had drafted the new Constitution. Even he was disillusioned with Aquino. 'We did not expect miracles,' he said, 'only a policy of damage limitation. But the fabric of society is falling apart.' I refrained from suggesting that unless the Philippines sewed shut the pockets of the greedy rich, it would cease to be a developing country and become a never-to-be-developed country. Aquino fell at the next election.

And most of the fishermen continue to dynamite the reefs and spray them with cyanide, although they know in their hearts that it will eventually result in the death of their livelihood, leaving only a haunted underwater graveyard.

Memento
Mari

Whitley Bay, Northumberland

I returned to Whitley Bay for the last time to attend my father's funeral. A tumour had spread inexorably over the pleural membrane that surrounded his lungs. It was so tough that a needle could not get through it to drain off the accumulating fluid. His lungs were being compressed by an internal tide that never ebbed.

Struggling to breathe was exhausting and he never strayed far from the large cylinder of oxygen that stood beside his chair. He sat gasping for air and worrying at the extent to which his once muscular calves had wasted away. When the World Cup was shown on television and he took no interest, I knew he had given up.

I visited the Garden of Remembrance where my mother's ashes had been scattered only a few years before. Then I wandered aimlessly through the town.

Whoever said that it is better to travel than to arrive had probably arrived here. Even in the least populated region of England, Whitley Bay seemed lifeless. I didn't meet anyone I knew, but then I was looking for my teenage schoolmates, not people who were now greying at the temples.

What little I remember of that day was inconsequential. Novelty shops, those treasure houses of tat. Comic postcards, familiar even after all these years, and still peopled by the fearsome, fat-bottomed 'missus' who was the butt of all jokes, and her puny henpecked husband who was forever searching for 'little Willie'. I also counted a couple of dozen charity shops, surely a bad sign.

Almost all Britain's seaside resorts have fallen on hard times. Their sad hotels now resemble residential homes. In Morecambe, whole streets of bed-and-breakfast guest houses are full of the homeless and are due to be compulsory purchased and demolished just to make sure the residents would be truly without homes. At night the seafront gloom is punctuated by strings of coloured lights, but every year there are fewer visitors and more dud bulbs. Some landlords now leave the distempering to the seagulls. One day, perhaps, the tide might go out and forget to return.

I wandered into the fairground that smelled of hamburgers and diesel. The fortune-teller's booth was closed and, unbelievably, the sign said, 'owing to unforeseen circumstances'. A dog on a string was tied to the railings while its master and his girl risked life and limb on the rickety Waltzer. Apart from two giggly girls with candyfloss stuck in their hair, they were the only passengers. The last waltz surely wasn't far away.

Was this really the Spanish City that had been celebrated in Dire Straits' 'Tunnel of Love', and where I had squandered my youth? It was now in its dotage and not long for this world. There were even rumours that the only significant local landmark, the Rotunda housing the Empress Ballroom, might be demolished.

At low tide I walked over the damp causeway to St Mary's Island where the lighthouse was soon to celebrate its centenary. The island was now a marine nature reserve; indeed, a third of the entire coastline of Britain is protected in some way and a third is ruined beyond redemption. I was told St Mary's Island attracted 100,000 visitors a year. It must have been a slack day or perhaps I was just alone no matter how many people were around.

There were no divers to be seen, although there were now 70,000 sport divers in Britain making one and a half million dives a year into our coastal waters. Perhaps below the deserted beaches of many resorts the submerged sands were crowded with unseen visitors, but probably not today at St Mary's Island. They were all in the Maldives or the Lesser Antilles jostling to be first to the reef.

I stared over the dark waters as the moon's tidal metronome ticked away the time. In my mind I swam out and dipped into one of the turbulent gullies. I followed the kelp-lined corridor into the past to when I had first dived here almost forty years before. It was of course exactly as I remembered, but when you trespass in time it is wise to be aware that, no matter how vivid the recollection, memory is an unreliable mechanism for remembering.

It was as if I had never been away, or perhaps I had been here in the future all the time, awaiting my arrival.

Nowadays, few coasts are so remote that they are unvisited by tourists. Anywhere there is warm water and a reef or wreck there are dive boats arriving every thirty minutes. While I don't begrudge the visitors their experience, I fear for the well-being of some sites.

Perhaps they will be spared thanks to youth's infatuation with hyper-realistic but unreal worlds. What if the young can't be bothered to dive or climb? Maybe it is too fanciful to imagine obese future generations who have evolved an enlarged mouse-clicking finger, but have useless atrophied legs dangling beneath their motorised computer chairs. Schools are even adopting 'virtual fieldwork' so that children can experience the outdoors without going outside. What a world they will miss.

The trouble is we take the sea for granted. There it lies at the edge of the land where it has always been. It doesn't do much, just comes in and goes out and gets agitated now and then. A succession of waves arrive only to be wrecked on the shore, yet still more come and it never occurs to us to ask why or from where.

We occasionally wade in the shallows and feel the undertow eroding the sand from beneath our feet, or we swim out a little and glance below at the sea floor. A few will even snorkel down for a closer look.

Perhaps having to go *down* into the sea is the problem. Down is unappealing, as in downtrodden, downcast, downhearted. Heaven is above, not below.

What if we could start again and discover the sea for the first time, a sea that was not horizontal but vertical? Imagine the splendour of an ocean reaching up into the sky, an undulating cliff of quicksilver moving as if it were alive. Extend a hand. It is engulfed and rapidly withdrawn, but returns unharmed, glistening and seasoned with salt.

If the sea were vertical, think of the trepidation and excitement of the first time you dared to walk through the wall of water to discover an alien landscape full of life, where not a single animal or plant resembled anything on dry land. You have changed planets.

Divers enjoy the best of both worlds. They regularly venture into a place where what we affectionately call the 'real world' is just a rumour. My first sight of the ascending medusae of my expelled breath not only reassured me that I was still alive in a place I was never designed to inhabit, but also revealed that air, the invisible stuff that had surrounded me throughout my life, was stunningly beautiful. You cannot fail to be entranced by the view through the surface film to that shimmering air-filled place above. The half-mirror, half-misty window that is the skin of the sea gives a magical view of a strange place you thought you knew well.

Once, when diving to get out of the rain, I noticed that someone was throwing gleaming pearls into the water. It was a just another mundane shower such as I had seen a thousand times before, but as I had never seen it until that moment.

*

If I have spent my life searching for the perfect shore, there were glimpses of perfection almost everywhere I looked. Even here. Gazing into the aquarium ocean of a tide pool on one of the saddest days of my life, the enthusiasm I have never managed to suppress rose again. Once you have a taste for the ocean, the intoxication lasts a lifetime.

Still, I envy those who explore this world for the very first time to discover the strung beads of magenta weed, hermit crabs with eyes on stalks and anemones with tentacles outstretched in anticipation of a lucky lunch. I wish tomorrow's children all the surprise and wonder that have been mine.

The things I've chosen are a drop, no more;
The undiminished sea still crowds the shore.

Ziya Pasha

Bibliography

Northumberland

Bousfield, E.L., and Le Blond, P.H., An account of *Cadborosaurus willsi*, a large aquatic reptile from the Pacific coast of North America, in *Amphipacifica* I: suppl. 3–25, 1995

British Association, *Official and Local Guide: Geology and Natural History*, Lambert & Co., Newcastle-upon-Tyne, 1889

Buckland, F., *Curiosities of Natural History*, 4th series, Richard Bentley & Son, London, 1873

Bullen, F., *The Cruise of the Cachalot*, Collins, London & Glasgow, 1898, repr. 1953

Churchill, W.S., *The Great War*, George Newnes Ltd, London, 1933

Corbin, A., *The Lure of the Sea*, Penguin Books, London, 1994

Earnshaw, T.S., *Hartley and Seaton Sluice: A Short History*, Robinson & Co. Ltd, Bedlington, 1961

Harris, B., *Sir John Vanburgh*, Longman, Green & Co., London, 1967

Heuvelmans, B., *In the Wake of Sea-Serpents*, Hill & Wang, New York, 1968

Howell, S., *The Seaside*, Studio Vista, Cassell & Collier Macmillan Publishers Ltd, London, 1974

King, H, 'Half-human creatures', in J. Cherry, *Mythical Beasts*, British Museum Press, London, 1995

Lencek, L., and Bosker, G., *The Beach*, Pimlico, London, 1999

McCanch, N., *A Lighthouse Notebook*, Michael Joseph, London, 1985

Marsden, C., *The English at the Seaside*, Collins, London, 1947

Mathews, L.H., 'The sperm whale, *Physeter catodon*', in *Discovery Reports* 17: 96–168, 1938

Mee, A., *Northumberland England's Furthest North*, Northumberland Education Committee, Newcastle-upon-Tyne, 1953

Milton, J., *Paradise Lost*, 1667, A. Fowler, ed., Longmans, London, 1979

Paxton, C.G.M., 'A cumulative species description curve for large open water marine animals', in *Journal of the Marine Biological Association of the United Kingdom* 78: 1389–91, 1998

Pevsner, N., *The Buildings of England: Northumberland*, Penguin Books Ltd, Harmondsworth, 1957

Pontoppidan, E., *Veruch einer Natürlichen Historie van Norwegen*, F.C. Mumme, Kopenhagen, 1753–4

Scott, P., and Rines, R., 'Naming the Loch Ness Monster', in *Nature* 258: 465–8, 1975

Steinbeck, J., *The Log from the Sea of Cortez*, Mandarin Books, London, 1951, repr. 1990

Trevelyan, G.M., *English Social History*, Longmans, Green & Co., London, 1944

Ilfracombe

Gatty, M., *British Seaweeds*, London, 1872

Gosse, E., *Father and Son*, Penguin Books Ltd, Harmondsworth, 1907, repr. 1949

Kingsley, C., *Glaucus, or the Wonders of the Shore*, Macmillan & Co., Cambridge, 1859

Lewes, G.H., *Sea-side studies at Ilfracombe, Tenby, the Scilly Isles and Jersey*, William Blackwood & Sons, Edinburgh and London, 1858

Liverpool

Gosse, E., *Father and Son*, Penguin Books Ltd, Harmondsworth, 1907, repr. 1949

Gosse, P.H., *Omphalos: An Attempt to Untie the Geological Knot*, John Van Voorst, London, 1857

Wales

Carefoot, T.H., Pennings, S.C., and Danko, J.P., 'A test of novel function(s) for the ink of sea hares', in *Journal of Experimental Marine Biology & Ecology* 234: 185–97, 1999

Cohen, J., 'Why so many sperms? An essay on the arithmetic of reproduction', *Science Progress* 57: 23–41, 1969

Fortey, R., *The Hidden Landscape*, Jonathan Cape, London, 1993

Gerard, J., *The Herball or Generall Historie of Plants*, enlarged and amended by T. Johnston, Islip, Norton & Whitaker's, London, 1633

Gosse, P.H., *A Year at the Shore*, Alexander Strahan, London, 1865

Herdman, W.A., 'The Liverpool Marine Biology Station on Puffin Island', in *Nature* 21 July 1887, 275–7

Herdman, W.A., *The foundation of the first season's work of the Liverpool Marine Biological Station on Puffin Island*, Turner, Routledge & Co., Liverpool, 1888

Lankester, R., *Diversions of a Naturalist*, Methuen & Co. Ltd, London, 1915

Morris, S.C., 'Once we were worms', in *New Scientist*, 2 August 2003, 34–7

Randerson, J., 'How we got our backbone', in *New Scientist*, 21 December 2002, 16

Rappoport, A.S., *The Sea – Myths and Legends*, Senate, London, 1995

Warner, M., *The Dragon Empress*, Weidenfeld & Nicolson, London, 1972

Canary Islands

Billet, D.S.M., 'Deep-sea holuthurians', in *Oceanography, Marine Biology Annual Review* 29: 259–318, 1991

Bory de St Vincent, J.B.G.M., *Essais sur les Isles Fortunées et L'Antique Atlantide*, Baudouin, Paris

Brown, A.S., *Brown's Madeira, Canary Islands and Azores*, Simpkin, Marshall, Hamilton, Kent & Co. Ltd, London, 1922

Butcher, R.W., 'Zostera: a report on the present condition of eelgrass on the coasts of England', in *Journal de Conseil Permanent Internationale pour L'exploration de la Mer* 9: 49–65, 1934

Diolé, P., *The Undersea Adventure*, Sidgwick & Jackson Ltd, London, 1953

Driscoll, E.M., Hendry, G.L., and Tinkler, K.J., 'The geology and geomorphology of Los Ajaches, Lanzarote', in *Geological Journal* 4: 321–334, 1965

Evans, H.M., *Sting-fish and Seafarer*, Faber & Faber, London, 1943

Fort, T., *The Book of Eels*, HarperCollins, London, 2002

Gilpatric, G., *The Compleat Goggler*, Dodd, Mead & Co., Inc., New York, 1938

Hanlon, R.T., 'The functional organization of chromatophores and irridescent cells in the body patterning of *Loligo plei*', in *Malacologia* 23: 89–119, 1982

Hausen, H., 'On the geology of Lanzarote, Graciosa and the isletas', in *Societas Scientiarum Fennica Commentationes Physico-Mathematicae* 23 (4): 1–116, 1959

Mercer, J., *The Canary Islanders, their Prehistory, Conquest and Survival*, Rex Collings, London, 1980

Myhill, H., *The Canary Islands*, Faber & Faber, London, 1968

Pain, S., 'Musical pears', in *New Scientist*, 16 October 1999, 52–3

Short, F.T., Mathieson, A.C., and Nelson, J.I., 'Recurrence of the eelgrass wasting disease at the border of New Hampshire and Maine, USA', in *Marine Ecology Progress Series* 20: 89–92, 1986

Short, F.T., Muelstein, L.K., and Porter, D., 'Eelgrass wasting disease: cause and recurrence of a marine epidemic', in *Biological Bulletin* 173: 557–62, 1987

Stephen, A.C., and Edmonds, S.J., *The Phyla Sipuncula and Echiura*, British Museum, London, 1972

Stewart, I., 'Playing by the rules', in *Independent on Sunday* magazine, 14 June 1998, 68–71

Vroom, P.S., and Smith, C.M., 'Life without cells', in *Biologist* 50: 222–6, 2003

Wells, M.J., *The Octopus*, Chapman & Hall, London, 1978

Wilkens, H., Parzefall, J., and Lliffe, A., 'Origin of the marine stygofauna of Lanzarote', in *Mitteilungen aus dem Hamburgischen Zoologischen Museum und Institut* 83: 223–30, 1986

Wilkens, H., Parzefall, J., and Ribowski, A., 'Population biology and larvae of the anchialine crab *Munidopsis polymorpha* (Galatheidae) from Lanzarote', in *Journal of Crustacean Biology* 10: 667–75, 1990

Winter, C.S., and Marrow, P., 'Digital organisms', in *Biologist* 45: 111–14, 1998

Isle of Man

Anon., *Manx Sea Fishing 1600–1900s*, Manx Heritage Foundation & Manx National Heritage, Douglas, 1991

Bennett, A., *Anna of the Five Towns*, Methuen, London, 1902

Bil-Jentzsch, A., *Internment of women on the Isle of Man*, privately printed, 1998

Bruce, J.R., 'The internment camp and the Biological Station', in *Report of the Marine Biological Station at Port Erin, Isle of Man* 54–7: 23–5, 1946

Burrows, E.M., and Lodge, S.M., 'Note on the inter-relationships between *Patella, Balanus* and *Fucus* on a semi-exposed coast', in *Report of the Marine Biological Station at Port Erin, Isle of Man* 62: 30–4, 1950

Chappell, C., *Island of Barbed Wire*, Corgi Books, London, 1986

Copely, J., 'Ooze cruise', in *New Scientist*, 11 March 2000, 27–9

Davies M.S., and Hawkins, S.J., 'Mucus from marine molluscs', in *Advances in Marine Biology* 34: 2–73, 1998

Dobbs, H.E., *Follow a Dolphin*, Fontana/Collins, London, 1979

Fraser, M., *In Praise of Manxland*, Methuen & Co. Ltd, London, 1948

Forbes, E., *A History of British Starfishes*, John Van Voorst, London, 1841

Gray, J., 'Women and History', MA thesis, De Montfort University, 1997

Herdman, W.A., *Founders of Oceanography and Their Work*, Edward Arnold & Co., London, 1923

Jenkins, S.R., Beukers-Stewart, B.D., and Brand, A.R., 'Impact of scallop dredging on benthic megafauna: a comparison of damage levels in captured and non-captured organisms', in *Marine Ecology Progress Series* 215: 297–301, 2001

Jones, N.S., 'Browsing of *Patella*', in *Nature* 158: 557, 1946

Jones, N.S., 'Observations and experiments on the biology of *Patella vulgata* at Port St Mary, Isle of Man', in *Proceedings & Transactions of the Liverpool Biological Society* 56: 60–77, 1948

Jones, N.S., and Kain, J.M., 'Subtidal algal colonization following the removal of *Echinus*', in *Helgolander Wissenschaft Meeresuntersuchungen* 15: 460–66, 1967

Moore, G., and Jennings, S., (eds), *Commercial Fishing: the wider ecological impacts*, British Ecological Society, 2000

Osborne, R., *The Floating Egg*, Pimlico, London, 1999

Quammen, D,. *The Song of the Dodo*, Pimlico, London, 1997

Rehbock, P.F., *The Philosophical Naturalists*, University of Wisconsin Press, 1983

Wallace, A.R., 'On the tendency of varieties to depart indefinitely from the original type', in *Journal of the Proceedings of the Linnean Society (Zoology)* 3: 53–62, 1858

Wilson, G., and Geikie, A., *Memoir of Edward Forbes*, Macmillan & Co., Cambridge, 1861

Wyville Thompson, C., *The Depths of the Sea*, Macmillan & Co., London, 1873

Scotland

Allen, J.A., Barnett, P.R.O., Boyd, J.M., Kirkwood, R.C., Mackay, D.W., and Smith, J.C., 'The environment of the estuary and Firth of Clyde', in *Proceedings of the Royal Society of Edinburgh*, Section B, 90: 1–539, 1986

Anon., 'Lord Strathcona and Mount Royal, 1820–1914', in *Proceedings of the Royal Society of London*, 1915

Boyd, J.M., and Boyd, I.L., *The Hebrides: The Mosaic of the Islands*, Birlinn Ltd, Edinburgh, 1996

Bray, E., *The Discovery of the Hebrides*, Birlinn Ltd, Edinburgh, 1996

Churchill, W.S., *A History of the English-Speaking Peoples*, Vol. II, Cassell & Co. Ltd, London, 1956

Constance, A., *The Impenetrable Sea*, Oldbourne Book Co. Ltd, London, 1958

Cousteau, J-Y, 'At home in the sea', in *National Geographic*, 465–507, April 1964

Crick, B., *George Orwell: A Life*, Secker & Warburg, London, 1980

Earle, S.A., *Sea Change*, G.P. Putnam's Sons, New York, 1995

Farre, R., *Seal Morning*, Hutchinson, London, 1957

Gjevdik, B., Moe, H., and Ommundsen, A., 'Sources of the Maelströms', in *Nature* 388: 837–8, 1997

Homer, *The Odyssey*, Everyman's Library Edition, J.M. Dent & Sons, London, 1953

Howell, S., *The Seaside*, Studio Vista, Cassell & Collier Macmillan Ltd, London, 1974

Kilbracken, Lord, 'The long, deep dive', in *National Geographic*, May 1963, 718–31

Kitching, J.A., 'Studies in sublittoral ecology III: *Laminaria* forest on the west coast of Scotland; a study of zonation in relation to wave action and illumination', in *Biological Bulletin* 80: 324–37, 1941

Masters, D., *When Ships Go Down*, Eyre & Spottiswoode Ltd, London, 1932

Norton, T.A., 'The zonation of seaweeds on rocky shores', in G. Moore &

R. Seed eds, *The Ecology of Rocky Coasts*, Hodder & Stoughton, London, 1985

Osborne, B.D., and Armstrong, R., *Mungo's City: A Glasgow Anthology*, Birlinn Ltd, Edinburgh, 1999

Pain, S., 'Whisky galore', in *New Scientist*, 31 August 2002, 48–9

Pearcy, R.W., 'Sunflecks and photosynthesis in plant canopies', in *Annual Review of Plant Physiology & Plant Molecular Biology* 41: 421–53, 1990

Pennant, T., *A Tour in Scotland and Voyage to the Hebrides*, 1774 & 1776, Birlinn Ltd, Edinburgh, repr. 1998

Poe, E.A., 'A Descent into the Maelström', Nicholson & Watson, London, 1958

Prebble, J., *The Highland Clearances*, Penguin Books Ltd, Harmondsworth, 1969

Prebble, J., *Scotland*, Secker & Warburg, London, 1984

Ritvo, H., *The Platypus and the Mermaid*, Harvard University Press, 1997

Stephenson, T.A., and Stephenson, A., *Life Between Tidemarks on Rocky Shores*, W.H. Freeman & Co., San Francisco, 1972

Thompson, D., *The People of the Sea*, Barrie & Rockliff, London, 1965

Verne, J., *Twenty Thousand Leagues Under the Sea or The Marvellous and Exciting Adventure of Pierre Aronnax, Conseil his Servant, and Ned Land, a Canadian Harpooner*, G.M. Smith, Boston, 1875

Wright, J., *Encyclopedia of Sunken Treasure*, Michael O'Mara Books Ltd, London, 1995

Nowhere & Somewhere in the past

Anon., 'Dumbo goes diving', in *New Scientist*, 8 September 2001, 17

Douglas, K., 'Our distant ancestor's fondness for a swim may explain why humans are such unusual primates', in *New Scientist*, 23 November 2000, 29–33

Hardy, A., 'Was man more aquatic in the past?', in *New Scientist*, 17 March 1960, 642–5

Hong, S.K., and Rahn, H., 'The diving women of Korea and Japan', in *Scientific American* 216 (5): 34–43, 1967

Kooyman, G.L., 'Diving physiology', in W.E. Perrin, B. Würsig and J.G.M. Thewissen, eds, *Encyclopedia of Marine Mammals*, Academic Press, San Diego, 2002

Le Page, M., 'Can you train your eyes to see better underwater?', in *New Scientist*, 17 May 2003, 14

Maraini, F., *Hekura: The Diving Girls' Island*, Hamish Hamilton, London, 1962

Marden, L., 'Ama, sea nymphs of Japan', in *National Geographic*, July 1971, 122–35

Melville, H., *Typee or A Peep at Polynesian Life*, Murray, London, 1847

Milsom, B., 'Breathless – by choice', in *Biologist* 47 (5): 239–43, 2000

Morgan, E., *The Descent of Woman*, revised edn, Souvenir Press, London, 1989

Norton, T., *Stars Beneath the Sea*, Arrow, London, 2000

O'Brien, B., 'Cool fin Tanya', in *Dive*, April 1999, 32–6

Phillips, H., 'Into the abyss', in *New Scientist*, 31 March 2001

Richards, M.P., Schulting, R.J., and Hedges, R.E.M., 'Sharp shift in diet at onset of Neolithic', in *Nature* 425: 366, 2003

Roberts, E., 'Song of the whale falls on deaf ears', in *Marine Scientist* 4 (3): 36–8, 2003

Roede, M., Wind, J., Patrick, J.M., and Reynolds, V., eds, *The Aquatic Ape: Fact or Fiction?*, Souvenir Press Ltd, London, 1991

Taylor, M.A., 'Origins of marine mammals', in W.F. Perrin, B. Würsig and J.G.M. Thewissen, eds, *Encyclopedia of Marine Mammals*, Academic Press, New York & London, 2002

Ireland

Dayton, P.K., 'Competition, disturbance and community organization: The provision and subsequent utilization of space in a rocky intertidal community', in *Ecological Monographs* 41: 351–89, 1971

Ebling, F.J., 'The exploration of the Rapids', in A.A. Myers, C. Little, M.J. Costello and J.C. Partridge, eds, *The Ecology of Lough Ine*, Royal Irish Academy, Dublin, 1991

Firth, R., *Malay fishermen: their peasant economy*, Routledge & Kegan Paul Ltd, London, 1946

Forbes, E., *The Natural History of the European Seas*, John Van Voorst, London, 1859

Marchant, J., 'First Light', in *New Scientist*, 22 July 2000, 34–5

Meyer-Rochow, V.B., 'Light of my life – messages in the dark', in *Biologist* 48 (4): 163–67, 2001

Moulton, J.A., 'Underwater sound: Biological aspects', in *Oceanography, Marine Biology Annual Review* 2: 425–54, 1964

Norton, T., *Reflections on a Summer Sea*, Arrow, London, 2002

O'Faolain, S., *An Irish Journey*, Readers Union & Longmans Green, London, 1941

Paine, R.S., 'Marine rocky shores and community ecology: an experimentalist's perspective', in *Excellence in Ecology* 4: Ecology Institute, Oldendorf/Luhe, 1994

Pain, S., 'Squawk, burble and pop', in *New Scientist*, 8 April 2000, 42–5

Renouf, L.P.W., and Rees, T.K., 'A note on experiments concerned with biotic factors on the sea-shore', in *Annals of Botany* 46: 1061–2, 1932

Tracy, H., *Mind You, I've Said Nothing*, Methuen, London, 1953

Wolman, D., 'Calls from the deep', in *New Scientist*, 15 June 2002, 35–7

Washington

Graves, W., 'The sea gate of the Pacific Northwest, Puget Sound', in *National Geographic*, January 1977, 71–97

Nordhoff, C., *Nordhoff's West Coast: California, Oregon and Hawaii*, 1874–5, KPI Ltd, London & New York, repr. 1987

Ohanian, B., 'Living a dream on the islands of the Puget Sound', in *National Geographic*, June 1995, 106–30

Raban, J., *Passage to Juno*, Picador, London, 1999

Ribera, M.A., and Boudouresque, C-F., 'Introduced marine plants, with special reference to macroalgae: mechanisms and impact', in F.E. Round and D.J. Chapman, *Progress in Phycological Research* 11: 187–268, 1995

Samson, J., '"That extensive": HMS *Herald*'s North Pacific survey, 1845–51' in *Pacific Science* 52: 287–93, 1998

Sculthorpe, C.D., *The Biology of Aquatic Vascular Plants*, Arnold, London, 1967

California

Arden, H., 'East of Eden: California's enterprise mid coast', in *National Geographic*, April 1984, 424–64

Astro, R., *John Steinbeck and Edward Ricketts: The Shaping of a Novelist*, University of Minnesota Press, 1973

Beebe, W., *Beneath Tropic Seas*, G.P. Putnam's Sons, New York and London, 1928

Cheng, L., '*Halobates*', in *Oceanography, Marine Biology Annual Review* 11: 223–7, 1973

Cheng, L., 'Marine pleuston – animals at the sea-air interface', in *Oceanography, Marine Biology Annual Review* 13: 181–212, 1975

Côte, I.M., 'Evolution and ecology of cleaning symbiosis in the sea', in *Oceanography, Marine Biology Annual Review* 38: 311–55, 2000

Culley, M., *The Pilchard: Biology and Exploitation*, Pergamon Press, London, 1971

Darwin, C., *Journal of Researches into the Natural History and Geology of the Countries Visited During the Voyage of H.M.S. Beagle Under the Command of Captain Fitzroy, R.N.*, John Murray, London, 1845

Dawson, E.Y., Neushul, M., and Wildman, R.D., 'Seaweeds associated with kelp beds along southern California and northwestern Mexico', in *Pacific Naturalist* 1: 1–81, 1960

Dugan, J., *Man Explores the Sea*, Penguin Books, Harmondsworth, 1960

Foster, M.S., and Shiel, D.R., 'The ecology of giant kelp forests in California', in *Fish & Wildlife Service, U.S. Department of the Interior, Biological Report* 85, 1985

Hanauer, E., *Diving Pioneers: An Oral History of Diving in America*, Watersport Publishing Inc., San Diego, 1994

Humphreys, J., 'Ed Ricketts and Cannery Row', in *Biologist* 40 (2): 62–6, 1993

Lewis, R.J., Norris, J.N., and Markham, J.W., 'In memoriam: Michael Neushul Jr, 1933–1993', in *Applied Phycology Forum* 10: 5–12, 1993

Limbaugh, C., 'Fish life in the kelp and the effects of kelp harvesting', in *University of California Institute of Marine Resources* IMR ref. 55–9, 1955

Limbaugh, C., 'Cleaning symbiosis', in *Scientific American* 205: 42–9, 1961

Minshall, R.L., *Window on the Sea*, Copley Books, La Jolla, 1980

Neushul, M., 'Diving in Antarctic waters', in *The Polar Record* 10: 352–8, 1961

Neushul, P., and Wang, Z., 'Between the devil and the deep sea: C.K. Tseng, mariculture, and the politics of science in modern China', in *Isis* 91: 59–89, 2000

North, W.J., 'The last dive', in *Skin Diver*, July 1960, 26–8

North, W.J., *The Biology of Giant Kelp Beds*, Verlag von J. Kramer, Lehre, 1971

North, W.J., 'Giant kelp, Sequoias of the sea', in *National Geographic* 142: 194–218, 1972

North, W.J., 'Connie Limbaugh's last dive', in E. Hanauer, *Diving Pioneers: An Oral History of Diving in America*, Watersport Publishing Inc., San Diego, 108–17, 1994

Norton, T.A., 'Gamete expulsion and release in *Sargassum muticum*', in *Botanica Marina* 24: 465–70, 1981

Palmer, J.D., *The Biological Rhythms and Clocks of Intertidal Animals*, Oxford University Press, 1995

Reed, D.C., Norris, J.N., and Foster, M.S., 'Dr Michael Neushul, Jr', in *Botanica Marina* 37: 287–92, 1994

Ricketts, E.F., and Calvin, J., *Between Pacific Tides*, 3rd edn revised by J. Hedgepeth, Stanford University Press, 1952

Simpson, D.A., 'One in a million', in *Aquarium Journal*, May, 1962, 185–9

Spies, B., and Davies, P.H., 'The infaunal benthos of a natural oil seep in the Santa Barbara channel', in *Marine Biology* 50: 227–37, 1979

Steinbeck, J., *Cannery Row*, 1945, repr. in *The Short Novels of John Steinbeck*, The Companion Book Club, London, 1956

Steinbeck, J., 'About Ed Ricketts', in *The Log from the Sea Of Cortez*, 1951, repr. Mandarin Books, London, 1990

Steinbeck, J., and Ricketts, E.F., *Sea of Cortez: A Leisurely Journal of Travel and Research*, The Viking Press, New York, 1941

Vermeij, G.J., and Dudley, R., 'Why are there so few evolutionary transitions between aquatic and terrestrial ecosystems?', in *Biological Journal of the Linnean Society* 70: 541–54, 2000

Zeng C., Abello, P., and Naylor, E., 'Endogenous tidal and semilunar moulting rhythms in early juvenile shore crabs *Carcinus maenus*', in *Marine Ecology Progress Series* 191: 257–66, 1999

Sweden

Burkholder, J., 'Scary monster super creeps', in *New Scientist*, 3 June 2000, 42–5

Hall, A., and Stephenson, C., 'Phocine distemper epidemic', in *Planet Earth*, winter 2002, 22–3

Hallegraeff, G.M., 'A review of harmful algal blooms and their apparent global increase', in *Phycologia* 32: 79–99, 1993

Harwood, J., 'How individual variation can affect the population dynamics of marine mammals', in *Scottish Association for Marine Science Newsletter* 19: 9–10, 1998

Harwood, J., 'Lessons from the seal epidemic', in *New Scientist*, 18 February 1989, 38–42

Mackenzie, D., 'Sick to death', in *New Scientist*, 5 August 2000, 33–5

Richardson, K., 'Harmful or exceptional phytoplankton blooms in the marine ecosystem', in *Advances in Marine Biology* 31: 302–85, 1997

Trainer, V.L., 'Unveiling an ocean phantom', in *Nature* 418: 925–6, 2002

Yemen

Cockburn, P., 'They chant of bitterness and hell', in *Independent on Sunday*, 31 January 1999

Hourani, A., *A History of the Arab Peoples*, Faber & Faber, London, 1991

Mackintosh-Smith, T., *Yemen: Travels in Dictionary Land*, John Murray, London, 1997

Mao, S-H., and Chen, B-Y., *Sea Snakes of Taiwan*, National Science Council of Republic of China, Taiwan, 1980

Marco Polo, *The Travels of Marco Polo*, revised edn, Jonathan Cape, London, 1928

Marshall, A., 'Warring tribes may tear Arabia Felix apart', in *Independent on Sunday*, 8 May 1994

Smith, M., *Monograph of the Sea Snakes*, British Museum, London, 1926

Stark, F., *The Southern Gates of Arabia*, 1936, repr. Arrow Books, London, 1990

Egypt

Abercrombie, T.J., 'Egypt: Change comes to a changeless land', in *National Geographic*, March 1977, 312–43

Douglas, K., 'Reef encounter', in *New Scientist*, 29 July 2000, 26–30

Hawkins, J.P., and Roberts, C.M., 'Effects of recreational SCUBA diving on the fore-reef slope communities of coral reefs', in *Biological Conservation* 62: 171–8, 1992

Hawkins, J.P., and Roberts, C.M., 'Growth of coastal tourism in the Red Sea: fore-reef slope communities of coral reefs', in *Ambio* 23: 503–8, 1994

Hawkins, J.P., Roberts, C.M., and Adamson, T., 'Effects of a phosphate ship grounding on a Red Sea coral reef', in *Marine Pollution Bulletin* 22: 538–42, 1991

Madl, P., 'The phenomenon of Lessepsian migration', in A. Goldschmid, ed., *Colloquial Meeting of Marine Biology*, Salzburg, 1999

Ormond, R., and Douglas, A., eds, *The Exploitation of Coral Reefs*, Ecological Issues, British Ecological Society, 1996

Por, F.D., *Lessepsian Migration*, Springer-Verlag, Berlin, 1978

Roberts, C.M., and Hawkins, J.P., 'Oil spill at Sharm el Sheikh', *Marine Pollution Bulletin* 19: 92–3, 1988

Schonfield, H.J., *The Suez Canal*, Penguin Books, Harmondsworth, 1939

Various, 'Aspects of Egypt', in *Observer Travel Extra*, 17 November 1985

Walker, M,. 'She's gotta have it', in *New Scientist*, 29 January 2000, 22–6

Ward, F., 'Florida's coral reefs are imperilled', in *National Geographic*, July 1990, 115–32

Philippines

Beebe, W., *Beneath Tropic Seas*, G.P. Putnam's Sons, New York and London, 1928

Brown, E.B., 'Adaptations of reef corals to physical environmental stress', in *Advances in Marine Biology* 31: 221–99, 1997

Chadwick, D.H., 'Coral in Peril', in *National Geographic*, January 1999, 30–7

Davis, L., *The Philippines: People, Poverty and Politics*, Macmillan Press, London, 1987

Fenton, J., 'The snap revolution', in *Granta* 18: 33–148, 1986

Hamilton-Paterson, J., *Seven-tenths: The Sea and its Thresholds*, Hutchinson, London, 1992

Karnow, S., *In Our Image: America's Empire in the Philippines*, Ballantine Books, New York, 1990

Kirkup, J., *Filipinescas*, Phoenix House, London, 1968

Nowak, R., 'Great Barrier bluff', in *New Scientist*, January 6 2003, 8–10

Pearce, F., 'A greyer shade of green', in *New Scientist*, 21 June 2003, 40–3

Saeger, J., 'The Samar Sea, Philippines: a decade of devastation', in *NAGA: The ICLARM Quarterly*, October 1993, 4–6

Schirmer, D.B., and Shalom, S.R., *The Philippine Reader*, Ken Inc., Quezon City, 1987

Stalker, P., 'The Philippines under fire', in *New Internationalist* 205: 4–25, 1990

Wood, E., 'Bleaching five years on are the tables turning?', in *Marine Conservation* 6 (1): 28–9, 2003

Stars Beneath the Sea

Trevor Norton

This is a story of some of the brave, brilliant and often barmy men that invented diving. It is a story of explosive tempers and exploding teeth, of how to juggle live hand grenades and steer a giant rubber octopus.

A series of vivid portraits reveal the eccentric exploits of these pioneers. They include Guy who held a world altitude record when only sixteen, wrote a film for Humphrey Bogart, invented snorkelling and loved his wife enough to shoot her. Roy wore a backet over his head and stole a coral reef. Bill wearied of fishing with dynamite and wrestling deadly snakes, so he sealed himself in a metal coffin to dangle half a mile beneath the ocean. Cameron, testing the bouncing bomb for dam busters, made a plastic ear for a dog, a false testicle for a stallion and invented a mantrap disguised as a lavatory. He ascended from a depth of 200 feet without breathing equipment to see if his lungs would burst, then studied the effects of underwater explosions by standing closer and closer until shattered by the blast.

The book also traces the evolution from spear fishermen to conversationalists, from treasure hunters to archaeologists, from photographers to philosophers. The sea is a secretive and seductive place and the author describes the magic and mystery of being beneath the waves.

'Norton writes with wit and a fine eye for the poetry in the scientific work' *Guardian*

arrow books

Reflections on a Summer Sea

Trevor Norton

This is the funny and touching story of a menagerie of eccentric and talented ecologists who, mainly as a hobby, spent forty summers at Lough Ine, a stunning marine lough in a corner of Ireland, where myths seep from the ground like will o' the wisps and, in one of the most unlikely projects in the history of science, were responsible for the reinvention of marine biology.

Among the stars of the book are the marine creatures that occupy the lake: sea urchins that won't dine unless they wear a hat, otters that steal experiments, and worms that will only mate by order of the moon. The creatures' eccentric behaviour is matched only by that of the ecologists themselves, whose antics and interactions with their Irish neighbours are all lovingly described with Norton's keen eye for both the wonderful and the absurd.

But for all its humour, the book is also a moving account of two ecologists who collaborated for forty years until their friendship came to a tragic end. The book brings together all the rich flavours of Ireland, the wonders of natural history and the magic of being a marine biologist just for the fun of it.

'A lovely book – Norton writes beautifully' *Sunday Express*

'*Reflections on a Summer Sea* stands apart, a thoughtful, funny look at life as it was' *Home & Country*

'Norton captures wonderfully the wit of the local Irish neighbours – making me smile, laugh and scowl' *Dive Magazine*

arrow books

Order further Trevor Norton titles
from your local bookshop, or have them delivered
direct to your door by Bookpost

☐ **Stars Beneath the Sea**   0 09 940509 1   £8.99
☐ **Reflections on a Summer Sea**
                           0 09 941616 6   £6.99

Free post and packing

Overseas customers allow £2 per paperback

Phone: 01624 677237

Post: Random House Books
c/o Bookpost, PO Box 29, Douglas, Isle of Man IM99 1BQ

Fax: 01624 670923

email: bookshop@enterprise.net

Cheques (payable to Bookpost) and credit cards accepted

Prices and availability subject to change without notice.
Allow 28 days for delivery.
When placing your order, please state if you do not wish to receive any
additional information.

www.randomhouse.co.uk/arrowbooks

arrow books